Lecture Notes in Mathematics

Edited by A. Dold and B. Eckmann

978

Julian Ławrynowicz

in cooperation with

Jan Krzyż

Quasiconformal Mappings in the Plane:

Parametrical Methods

Springer-Verlag
Berlin Heidelberg New York Tokyo 1983

Authors

Julian Ławrynowicz
Institute of Mathematics of the
Polish Academy of Sciences, Łodź Branch
Kilińskiego 86, PL-90-012 Łodź, Poland

in cooperation with

Jan Krzyż
Institute of Mathematics of the
Maria-Curie-Sklodowska University in Lublin
Nowotki 10, PL-20-031 Lublin, Poland

AMS Subject Classifications (1980): 30 C 60

ISBN 3-540-11989-2 Springer-Verlag Berlin Heidelberg New York Tokyo
ISBN 0-387-11989-2 Springer-Verlag New York Heidelberg Berlin Tokyo

Printed in Germany

Printing and binding: Beltz Offsetdruck, Hemsbach/Bergstr.
2146/3140-543210

FOREWORD

These lecture notes contain an exposition of analytic properties of quasiconformal mappings in the plane (Chapter I), a detailed and systematic study of the parametrical method with complete proofs (exploring these analytic properties and including results of the author), partly new and simplified (Chapter II), and a brief account of variational methods (Chapter III).

In contrast to the books by Lehto and Virtanen [1, 2] and by Ahlfors [5] the present author starts in Chapter I with defining the class of quasiconformal mappings as the closure of the Grötzsch class with respect to uniform convergence on compact subsets. Then an important part of this chapter is devoted to proving the fundamental theorem on existence and uniqueness of quasiconformal mappings with a preassigned complex dilatation, established in particular cases by Gauss [1], Lichtenstein [1] and Lavrentieff [2], and in the general case by Morrey [1]. The present author chooses the proof due to Bojarski [2] which is based a.o. on the fundamental results of Calderón and Zygmunt [2] connected with properties of the Hilbert transform and on reducing the problem in question to solving some linear integral equation.

Parametrical and variational methods belong to the most powerful research tools for extremal problems in the complex analysis.

The parametrical method for conformal mappings of the unit disc $\{z: |z| < 1\}$, initiated by Löwner [1], consists in fact of studying a partial differential equation for a function $w = f(z,t)$, $|z| < 1$, $0 \le t \le T$, whose solutions are homotopically contractible to the identity mapping within the class of conformal mappings of the unit disc and form in this class a dense subclass. This method was then extended to the case of doubly connected domains by Komatu [1] and Golusin [1]. With the help of this method it was possible to obtain a number of basic information on conformal mappings inaccessible for elementary methods. An analogous method for quasiconformal mappings of the unit disc onto itself was initiated by Shah Tao-shing [1] and then extended by other authors.

In Chapter II, concerned with the parametrical method, we concentrate on the results of Shah Tao-shing [1], Ahlfors and Bers [1], Ahlfors [5], Gehring and Reich [1], and the author. When considering more advanced and special topics as well as applications, a special attention is paid to the results of Reich and Strebel [1-3], Gehring and Reich [1], Kühnau [1-13], and Lehto [3-6]. These topics are mainly connected with Teichmüller mappings, quasiconformal mappings with preassigned boundary values, and conformal mappings with quasiconformal extensions.

Thanks to the books by Belinskiĭ [2], Kruškal [4, 5] and Schober [1], there is no necessity to describe the variational methods here in detail; they are only briefly reviewed. For conformal mappings they were initiated by Hadamard [1], Julia [1] and Courant [1], and then developed as a very effective research tool in the papers by Schiffer [1], Schaeffer and Spencer [1], Golusin [2], and other authors. Investigation of extremal properties of quasiconformal mappings, in particular a characterization of extremal mappings and their connection with quadratic differentials is a deep and unexpected discovery of Teichmüller [1], which initiated the development of more special variational methods for quasiconformal mappings (Belinskiĭ [1], Schiffer [2], and others).

In Chapter III, connected with variational methods, we concentrate on the results of Teichmüller [1, 2], Belinskiĭ [1], Schiffer [2], Renelt [1, 3], Schiffer and Schober [1, 2], Kruškal [1-8], Strebel [2-5], and Kühnau [1-15]. Section 26 gives a brief account of those aspects of the theory of extremal quasiconformal mappings and problems connected with the famous Teichmüller theorem (Teichmüller [1], Ahlfors [2]) which are necessary for full motivation of the Teichmüller mappings considered earlier in Sections 17 and 18. The concluding Section 27 indicates the importance of quasiconformal mappings, in particular of the analytic approach as well as of parametrical and variational methods, in electrical engineering.

In this place the author would like to thank Profs. L. V. Ahlfors, S. L. Kruškal, R. Kühnau, O. Lehto, A. Pfluger, and Dr. T. Iwaniec for reading various parts of the manuscript and critical remarks.

Łódź, Poland, November 1982 Julian Ławrynowicz

CONTENTS page

Contents

NOTATION AND ABBREVIATIONS

Throughout these lecture notes, unless otherwise specified, we are concerned with points and sets of the closed plane $\mathbb{E}$, the one point compactification of the finite plane $\mathbb{C}$ with the usual metric. The real line with the usual metric is denoted by $\mathbb{R}$. The corresponding sets are denoted by $\mathbb{E}$, $\mathbb{C}$, and $\mathbb{R}$ as well.

The difference of two sets E and E' is denoted by $E \setminus E'$, the closure of E by $\mathrm{cl}\,E$, the interior of E by $\mathrm{int}\,E$, while the boundary, topological and oriented (positively) with respect to E, by $\mathrm{fr}\,E$ and ∂E, respectively. In the latter definition we assume E to be a domain bounded by disjoint Jordan curves (in particular, a Jordan domain), or its closure. Under Jordan curve we mean a homeomorphic image of a circle, under Jordan arc – a homeomorphic image of an interval, i.e. of a connected subset of $\mathbb{R}$ which does not reduce to a point. The open and closed segments with end points a, b are denoted by $(a;b)$ and $[a;b]$, respectively, while under (a,b) we mean the ordered pair of a, b and – more generally – (a_n) denotes the sequence with terms $a_1, a_2, \ldots$ We also put

$$\Delta^r(s) = \{z\colon |z - s| \le r\}, \quad \Delta^r = \Delta^r(0), \quad \Delta = \Delta^1, \quad \Delta_r = \Delta \setminus \mathrm{int}\,\Delta^r.$$

The Lebesgue plane measure of a measurable set $E \subset \mathbb{C}$ is denoted by $|E|$, the outer Lebesgue linear measure of a set $I \subset \mathbb{R}$ – by $[\![I]\!]^*$, while its Lebesgue linear measure, if I is measurable, by $[\![I]\!]$. Thereafter under a (plane, linear) <u>measurable</u> set we always understand a set which is Lebesgue (plane, linear) measurable. The diameter of a set E is denoted by $\mathrm{dia}\,E$.

If f is a function defined on E, and $E' \subset E$, then $f[E']$ denotes the image of E' under f (in case of a functional the convention concerning ( ) and [ ] is inverse). If, in particular, f is an elementary function: exp, arg, etc., and $z \in E$, we write fz instead of $f(z)$ in case where it does not lead to misunderstanding. An analogous convention is applied to linear operators. The zero and identity functions, defined on a set E, are denoted by 0_E and id_E, respectively. In the case where $E = \mathbb{E}$, the subscript is omitted. We say that f, defined on E, satisfies a property if this property is satisfied for all $z \in E$.

If f and g are functions defined on E and E', respectively, where $E' \cap f[E] \neq \emptyset$, then the composite function defined on the preimage of $E' \cap f[E]$ under f is denoted by $g \circ f$, and for any z of this preimage we write $g \circ f(z)$ instead of $(g \circ f)(z)$. Further, $\check{f}$ denotes the inverse of f, if it exists, while $f^{-1} = 1/f$. Correspondingly, $\check{f}[E']$ denotes the preimage of a set E' under f even in the case where f is not invertible. The notation $\check{f}$ for the inverse of f is used e.g. in Behnke and Sommer [1] and is much more convenient for our purposes than f^{-1} which e.g. in the expression f_w^{-1} (differentiation with respect to w) may lead to misunderstanding.

The support of f, i.e. the set $\mathrm{cl}\{z: f(z) \neq 0\}$, is denoted, following Morrey [2], by spt f; the abbreviation supp is more popular, but it is longer and similar to sup. A mapping (i.e. a function) from E into $\mathbb{R}$, $\mathbb{C}$, or $\mathbb{E}$ is often denoted by $f: E \to \mathbb{R}$, $\mathbb{C}$, or $\mathbb{E}$, respectively. In the case where f is one-to-one, we always write $f: E \to E'$, $E' \subset \mathbb{R}$, $\mathbb{C}$, or $\mathbb{E}$, and this means that $f[E] = E'$. We also write $a_n \to a$ instead of $a_n \to a$ as $n \to +\infty$. If mappings $f: E \to \mathbb{E}$ and $f_n: E_n \to \mathbb{E}$, $n = 1, 2, \dots$, are such that for every compact subset E_0 of $E \setminus \{\infty\} \setminus \check{f}[\{\infty\}]$ there is an index k such that $E_0 \subset E_n$ for $n > k$ and $f_n | E_0 \to f | E_0$ uniformly, we say, following Saks and Zygmund [1], that (f_n) _tends to_ f _almost uniformly_ and write $f_n \rightrightarrows f$. Finally, if $f: E \to \mathbb{C}$ is measurable, we put

$$\|f\|_p = \left(\iint_E |f|^p dxdy \right)^{1/p}, \quad 1 \le p < +\infty,$$

$$\|f\|_\infty = \begin{cases} 0, & |E| = 0, \\ \inf\limits_{E'} \sup\limits_{z \in E \setminus E'} |f(z)|, & |E| > 0, \end{cases}$$

where the infimum is taken over all sets E' with $|E'| = 0$.

The expressions _if and only if_, _almost every[where]_, and _with respect to_ are abbreviated by _iff_, _a.e._, and _w.r.t._, respectively, while _qc_ [_qcty_] means _quasiconformal[ity]_ and ACL – absolutely continuous on lines.

I. BASIC CONCEPTS AND THEOREMS IN THE ANALYTIC THEORY OF QUASICONFORMAL MAPPINGS

1. The class of regular quasiconformal mappings and its closure

We begin with the notion of a sense-preserving homeomorphism. It is well known that if $f: \operatorname{cl} D \to E$ is a homeomorphism and D a Jordan domain, then $E = \operatorname{cl} D'$, where D' is Jordan, and $f|\operatorname{fr} D: \operatorname{fr} D \to \operatorname{fr} D'$. Now, let $f_1: \operatorname{fr} \Delta \to \operatorname{fr} D$ and $f_2: \operatorname{fr} \Delta \to \operatorname{fr} D$ be homeomorphisms such that $\arg \check{f}_2 \circ f_1 | \operatorname{fr} \Delta \setminus \{-1\}$ is an increasing function of $\arg z$, $z \in \operatorname{fr} \Delta \setminus \{-1\}$. Then, as it is well known, also $g_1 = f \circ f_1$ and $g_2 = f \circ f_2$ are homeomorphisms such that $\arg \check{g}_2 \circ g_1 | \operatorname{fr} \Delta \setminus \{-1\}$ is an increasing function of $\arg z$, $z \in \operatorname{fr} \Delta \setminus \{-1\}$. Hence f induces a mapping between the orientation of $\operatorname{fr} D$ and the orientation of $\operatorname{fr} D'$. If they both are positive or negative w.r.t. the corresponding domains, $f: D \to D'$ is said to be <u>sense-preserving</u>. More general, if $f: E \to E'$ is a homeomorphism between two sets, f is said to be <u>sense-preserving</u> if $f|D$ is sense-preserving for every D such that $\operatorname{cl} D \subset E$. We note (cf. e.g. Newman [1], p. 198) that if $f: E \to E'$ is a homeomorphism and E is either a domain or the closure of a Jordan domain and there is a domain D such that $\operatorname{cl} D \subset E$ and $f|D$ is sense-preserving, then so is f. We also note that if f, f_1, and f_2 are sense-preserving, so are $\check{f}$ and $f_2 \circ f_1$ provided that $f_2 \circ f_1$ makes sense.

Suppose now that E is either an open set or the closure of a Jordan domain. A mapping $f: E \to \mathbb{E}$ is said to be <u>differentiable</u> at $z_0 \in \operatorname{int} E$ $(z_0, f(z_0) \neq \infty)$ if

$$(1.1)\quad f(z) = f(z_0) + f_z(z_0)(z - z_0) + f_{\bar z}(z_0)(\bar z - \bar z_0) + o(z - z_0),$$

where $f_z = \frac{1}{2}(f_x - if_y)$, $f_{\bar z} = \frac{1}{2}(f_x + if_y)$, $x = \operatorname{re} z$, $y = \operatorname{im} z$. It is clear that

$$(1.2)\quad \overline{f_z} = \bar f_{\bar z}, \quad \overline{f_{\bar z}} = \bar f_z .$$

A mapping f is said to be <u>differentiable</u> at ∞ if f^*, defined by

$f^*(z) = f(1/z)$, is differentiable at 0, and differentiable at z_0 such that $f(z_0) = \infty$ if f^{**}, defined by $f^{**}(z) = 1/f(z)$, is differentiable at z_0. A mapping $f: E \to \mathbb{E}$ is said to be differentiable if it is differentiable at every $z_0 \in \operatorname{int} E$. The directional derivatives $f_{|\alpha}$ are defined by $f_{|\alpha}(z) = e^{-i\alpha}[f_t(z + te^{i\alpha})]_{t=0}$, where α is real and $t \geq 0$. The Jacobian of f will be denoted by J or J_f. In the case where $z = \infty$ or $f(z_0) = \infty$ for some $z_0 \in E$, we do not define J_f but only $\operatorname{sgn} J_f(z_0) = \operatorname{sgn} J_{f^{**}}(z_0)$. It is easily verified that if $f: E \to \mathbb{C}$, where $E \subset \mathbb{C}$, is differentiable, then

$$(1.3) \quad f_{|\alpha} = e^{-i\alpha}(f_x \cos\alpha + f_y \sin\alpha) = f_z + e^{-2i\alpha} f_{\bar{z}},$$

$$(1.4) \quad \max|f_{|\alpha}| = |f_z| + |f_{\bar{z}}|, \quad \min|f_{|\alpha}| = ||f_z| - |f_{\bar{z}}||,$$

$$(1.5) \quad J = |f_z|^2 - |f_{\bar{z}}|^2 = \operatorname{sgn} J \max_\alpha |f_{|\alpha}| \min_\alpha |f_{|\alpha}|.$$

If f is differentiable at $z_0 \in \operatorname{int} E$ and $\operatorname{sgn} J_f \neq 0$, f is called regular at z and z is called a regular point of f. A mapping f is called regular if it is regular at every $z \in \operatorname{int} E$. A regular C^1-homeomorphism is called a diffeomorphism; in the case of sets E, E' containing ∞ the definition of C^1-functions should be extended similarly to that of differentiability. Suppose now that $f: E \to E'$ is a homeomorphism and E is either a domain or the closure of a Jordan domain. Then, by Newman's result quoted above, if $\operatorname{sgn} J(z) = 1$ at some regular z, f is sense-preserving and, conversely, if f is sense-preserving, $\operatorname{sgn} J(z) = 1$ at any regular $z \in \operatorname{int} E$.

Suppose that f is a diffeomorphism and D a domain. The ratio

$$(1.6) \quad p(z) \equiv p_f(z) = \max_\alpha |f_{|\alpha}(z)| / \min_\alpha |f_{|\alpha}(z)|, \quad z \in D,$$

is called the dilatation of f at z. Clearly it is bounded on every compact subset of D and invariant under conformal mappings. The second conclusion enables us to extend the definition of p to the cases $z = \infty$ and $f(z) = \infty$ analogously as the definition of differentiability was extended. A sense-preserving diffeomorphism $f: D \to D'$, where D is a domain and $\sup p(z) \leq Q < +\infty$, is called a regular Q-quasiconformal (shortly: regular Q-qc) mapping. These mappings, but not their name, were first introduced by Grötzsch [1] (cf. also Lavrentieff [1]).

The very natural definition of Grötzsch has the disadvantage that the class of regular Q-qc mappings is not closed w.r.t. almost uniform convergence. We are thus led to the following definition due to Lehto and Virtanen ([1], p. 222, or [2], p. 211): a nonconstant mapping f: $D \to \mathbb{E}$, D being a domain, is said to be Q-quasiconformal, if there is a sequence of regular Q-qc mappings f_n: $D_n \to D'$ such that $f_n \rightrightarrows f$ and for a.e. z for which there exist finite partial derivatives $f_z(z)$, $f_{\bar{z}}(z)$ we have $f_{n\bar{z}}(z)/f_{nz}(z) \to f_{\bar{z}}(z)/f_z(z)$. It is clear that if, in particular, f is a homeomorphism it is sense-preserving. The fact that any qc mapping is a homeomorphism will be proved in Section 11.

Actually the above definition is not exactly the same as that given by Lehto and Virtanen since they suppose that $D_n = D$, but as noticed by Gehring [2] and follows from their book (pp. 79-82 or 74-78) it makes no difference. It is worth-while to mention that, as noticed by Gehring [2] and follows from the quoted book (pp. 217-222 or 207-211), the restriction concerning the partial derivatives is superfluous since if we find a sequence of regular Q-qc mappings $f_n \rightrightarrows f$, we can always find another sequence with the required property and even a sequence of real-analytic mappings with $(f_{n\bar{z}}/f_{nz})$ consisting of polynomials. Since the proofs of these results are rather long and at the same time the presentation in the book of Lehto and Virtanen [1, 2] is excellent, we prefer to leave these aspects aside, all the more that we never make any use of them. For the problem in question cf. also Strebel [6].

According to Gehring [2] this is the most natural definition of qcty. We complete it by the following: a homeomorphism f: $D \to D'$, where D is the closure of a domain bounded by disjoint Jordan curves (in particular, of a Jordan domain) is said to be Q-quasiconformal if $f|\operatorname{int} D$ is Q-qc. The same completion is accepted for conformal mappings.

It is clear that the class of 1-qc mappings is identical with the class of conformal mappings. Thus qc mappings are their natural generalization. In many results on conformal mappings only qcty is essential, and various extremal problems in qc mappings lead to conformal mappings. On the other hand qc mappings are less rigid than conformal mappings, so they are much more flexible as a tool (cf. Ahlfors [5], pp. 1-2).

2. Differentiability

We are going to give (in Section 11) an analytic characterization of qc mappings which involves in a more clear way differentiability properties. This characterization in its final form is due to Gehring and Lehto [1], but it was originated by Strebel [1] and Mori [1]. We shall often refer to Saks [2] (in the bibliography we also list Saks [1]).

We begin with the notion of ACL. A continuous function $f: D \to \mathbb{E}$ is said to be absolutely continuous on lines (shortly: ACL) if for any rectangle U, $\operatorname{cl} U \subset D \setminus \{\infty\} \setminus \check{f}[\{\infty\}]$, with sides parallel to the coordinate axes it is absolutely continuous on almost all line segments in U which are parallel to either side of U.

We claim that if f is Q-qc, it is ACL.

In fact, modelling an idea of Pfluger [2], take U and y so that $U \cap \{z\colon \operatorname{im} z = y\} \neq \emptyset$. The function σ, defined by $\sigma(y) = |f[U \cap \{z\colon \operatorname{im} z < y\}]|$, is an increasing function, whence there exists a finite derivative $\sigma'(y)$ for every y in question except perhaps for a set of linear measure zero. Consider now a sequence (f_n) of regular Q-qc mappings as in the definition of Q-qcty of f. We are going to show that there exists a subsequence (f_{n_j}) such that for a.e. y the corresponding $\sigma'_{n_j}(y) \to \sigma'(y)$ and next that any such y gives rise to f absolutely continuous for fixed y.

Obviously $\sigma_n - \sigma$ is of bounded variation and hence the integral of $|\sigma'_n - \sigma'|$ over any interval in the domain of $\sigma_n - \sigma$ does not exceed the total variation of $\sigma_n - \sigma$ in this interval which tends to 0 as $n \to \infty$ since $f_n \rightrightarrows f$. Therefore $|\sigma'_n - \sigma'| \to 0$ in measure (cf. e.g. Graves [1], p. 236), whence, by F. Riesz's theorem (e.g. ibid., p. 244), (σ'_n) contains a subsequence (σ'_{n_j}) such that $\sigma'_{n_j} \to \sigma'$ a.e. In what follows we may drop the index j.

Now let $(x_k; \tilde{x}_k)$, $k = 1, \ldots, m$, belong to $\{x\colon x + iy \in U\}$ and be disjoint. Clearly, for each k and $\tilde{y} > y$ we have, by (1.1),

$$(2.1) \quad (\tilde{y} - y)|f_n(\tilde{x}_k + iy_{k,n}) - f_n(x_k + iy_{k,n})| \leq \int_y^{\tilde{y}} \int_{x_k}^{\tilde{x}_k} (|f_{n\zeta}| + |f_{n\bar{\zeta}}|)\,d\xi\,d\eta,$$

where $\xi = \operatorname{re} \zeta$, $\eta = \operatorname{im} \zeta$, and $y_{k,n}$ is chosen so that the length of the image arc of $\{\xi + iy_{k,n}\colon x_k \leq \xi \leq \tilde{x}_k\}$ is a minimum. Owing to the Schwarz inequality and the fact that f_n are regular Q-qc mappings, the right-hand side of (2.1) squared is estimated from above by

$$\iint_{R_k} \frac{|f_{n\zeta}| + |f_{n\bar{\zeta}}|}{|f_{n\zeta}| - |f_{n\bar{\zeta}}|}\,d\xi\,d\eta \iint_{R_k} (|f_{n\zeta}|^2 - |f_{n\bar{\zeta}}|^2)\,d\xi\,d\eta$$
$$\leq Q(\tilde{x}_k - x_k)(\tilde{y} - y)|f_n[R_k]|,$$

where $R_k = \{\xi + i\eta\colon x_k \leq \xi \leq \tilde{x}_k,\ y \leq \eta \leq \tilde{y}\}$. Consequently, applying the Schwarz inequality for sums, we conclude that

$$\sum_{k=1}^{m} |f_n(\tilde{x}_k + iy_{k,n}) - f_n(x_k + iy_{k,n})| / [\sum_{k=1}^{m} (\tilde{x}_k - x_k)]^{\frac{1}{2}}$$

$$\leq [Q \frac{\sigma_n(\tilde{y}) - \sigma_n(y)}{\tilde{y} - y}]^{\frac{1}{2}}.$$

Hence, letting $\tilde{y} \to y$, we obtain

$$\sum_{k=1}^{m} |f_n(\tilde{x}_k + iy) - f_n(x_k + iy)| \leq [Q\sigma_n'(y)]^{\frac{1}{2}} [\sum_{k=1}^{m} (\tilde{x}_k - x_k)]^{\frac{1}{2}}$$

and an analogous inequality for f, what means that f is ACL, as desired, since in the above consideration the roles of x and y can be interchanged.

Now, since f is ACL, it is of bounded variation on a.e. line segment in U which is parallel to either side of U. We claim that f has finite partial derivatives f_x and f_y a.e. in D. Indeed, if U^* is the set of all points of U, where f_x exists, and χ_{U^*} is its characteristic function, then χ_{U^*} is measurable as U^* is a Borel set, so, by the theorem of Fubini, we have

$$|U^*| = \iint_U \chi_{U^*} \, dxdy = \int_{y: x+iy \in U} \| U^* \cap \{z: \operatorname{im} z = y\} \| \, dy.$$

But f_x exists a.e. on $U \cap \{z: \operatorname{im} z = y\}$ for a.e. $y \in \{y: x + iy \in U\}$, whence $|U^*| = |U|$. Consequently f_x exists a.e. in D and, clearly, the same is true for f_y, as desired. Hence also f_z and $f_{\bar{z}}$ exist a. e. in D. Furthermore, (1.4) and (1.6) with f_n substituted for f as well as the definition of Q-qcty yield

$$(2.2) \quad |f_z| + |f_{\bar{z}}| \leq Q(|f_z| - |f_{\bar{z}}|) \text{ a.e.}$$

Therefore we have proved

<u>LEMMA 1</u>. <u>If</u> $f: D \to E$ <u>is</u> Q-<u>qc</u>, <u>it is</u> ACL <u>and possesses a.e. partial derivatives which satisfy</u> (2.2).

In Section 11 we shall prove that the converse is also true, provided f is a sense-preserving homeomorphism and D a domain.

3. Distributional derivatives

Let $f: D \to \mathbb{C}$ be a locally integrable function in a domain $D \subset \mathbb{C}$. Functions $f_z', f_{\bar{z}}': D \to \mathbb{C}$ locally L^p (i.e. functions which are L^p on arbitrary compact subsets of D) are said to be distributional L^p-derivatives of f, if

(3.1) $$\iint_D f'_z h\,dxdy = -\iint_D f\,h_z\,dxdy, \quad \iint_D f'_{\bar z} h\,dxdy = -\iint_D f\,h_{\bar z}\,dxdy$$

for all C^1-functions $h: D \to \mathbb{C}$ with compact support. When considering the classes L^p we always assume that $p \geq 1$. This definition can immediately be generalized to the case $D, f[D] \subset \mathbb{E}$, assuming the above conditions for D replaced with $D \setminus \{\infty\} \setminus \check{f}[\{\infty\}]$. It is clear that the idea of defining the distributional L^p-derivatives with the help of (3.1) originates from Green's formula which, in its classical version, yields

(3.2) $$\iint_{D^*} h^*_z\,dxdy = \tfrac{1}{2}i \int_{\partial D^*} h^* d\bar z, \quad \iint_{D^*} h^*_{\bar z}\,dxdy = -\tfrac{1}{2}i \int_{\partial D^*} h^* dz,$$

where D^* is a domain in $\mathbb{C}$ with rectifiable ∂D^* consisting of disjoint Jordan curves and $h^*: \mathrm{cl}\,D^* \to \mathbb{C}$ is a C^1-function.

We are going to show that any qc mapping f possesses distributional L^p-derivatives f'_z, $f'_{\bar z}$, and $f'_z = f_z$, $f'_{\bar z} = f_{\bar z}$ a.e. We begin with a lemma due to Gehring and Lehto [1].

By an open mapping we mean such a mapping $f: D \to \mathbb{E}$ that if $U \subset D$ and U is open, $f[U]$ is open as well. Under a Borel function we understand a function $f: E \to \mathbb{R}, \mathbb{C}$, resp. $\mathbb{E}$ such that the preimage $\check{f}[U]$ of each open set $U \subset \mathbb{R}, \mathbb{C}$, resp. $\mathbb{E}$ is Borel. A point z_0 is said to be a point of linear density for a set E in the x (resp. y) direction if for the collection of closed line segments I containing z_0 and parallel to the x (resp. y) axis we have $[\![E \cap I]\!]^*/[\![I]\!] \to 1$ as $[\![I]\!] \to 0$.

LEMMA 2. *If $f: D \to \mathbb{C}$, D being a domain in $\mathbb{C}$, is continuous and open, and possesses finite partial derivatives a.e. in D, it is differentiable a.e. in D.*

Proof. It is sufficient to prove the assertion for a compact $U \subset D$. We claim that for any $\eta > 0$ there is a closed set $U \subset D$ such that $|U \setminus E| < \eta$ and $f_x|E$, $f_y|E$ are continuous. In fact, let D^* denote the set where the both partial derivatives exist and are finite. For every $z \in D^*$ we define

(3.3) $$f(z,t) = \left| \frac{f(z+t) - f(z)}{t} - f_x(z) \right| + \left| \frac{f(z+it) - f(z)}{t} - f_y(z) \right|,$$

where $t \in \mathbb{R} \setminus \{0\}$ and $z+t,\ z+it \in D$. For sufficiently small $|t|$, the functions $f(\ ,t)$ and, consequently, the supremum g_n of them taken, at a fixed z, over all rational $t \in (0;1/n)$, n being sufficiently large, are defined a.e. in U and Borel. Since $g_n \to 0$ a.e. in D, we can apply Egoroff's theorem (cf. e.g. Saks [2], p. 18), and this yields the existence of a closed set $E \subset U$ such that $|U \setminus E| < \eta$ and the functions of z and t, defined on E at a given t by $(1/t)[f(z+t)-f(z)]$ and $(1/t)[f(z+it)-f(z)]$, both tend uniformly to $f_x|E$ and $f_y|E$, respectively. Hence the latter two functions are continuous, as desired. This means that for any $\varepsilon > 0$ there is a $\delta > 0$ such that $\max(|z^*-z_o|, |t|) < \delta$ implies

(3.4) $\max\{|f(z,t)|,\ |f_x(z^*)-f_x(z_o)|,\ |f_y(z^*)-f_y(z_o)|\} < \varepsilon;\ z_o, z^*, z \in E.$

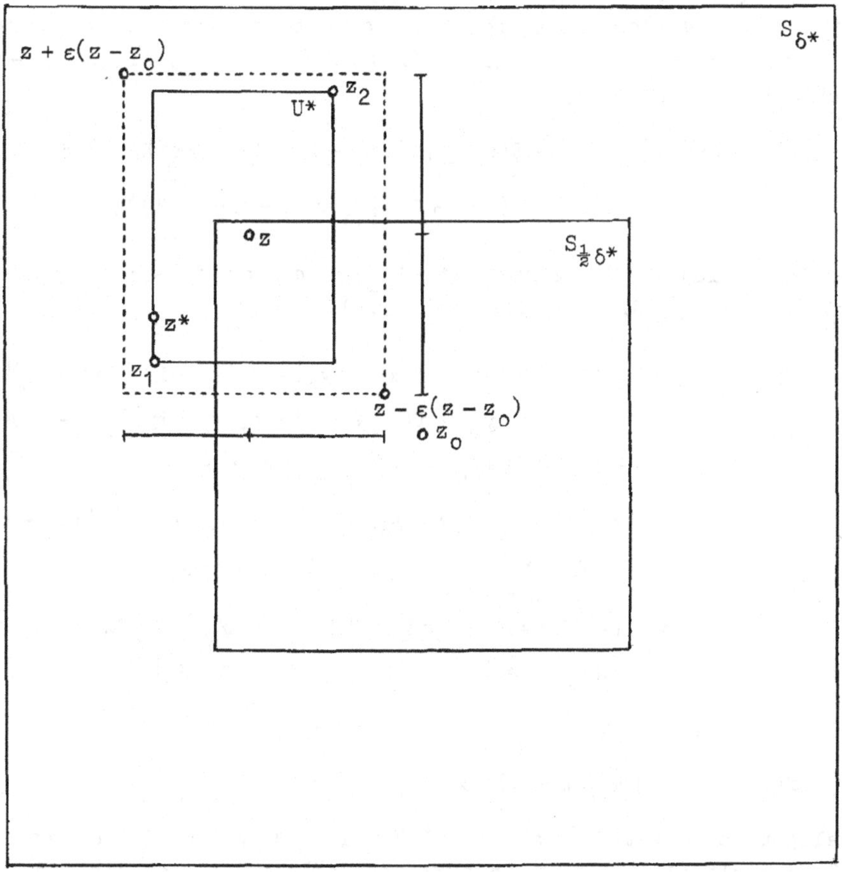

Fig. 1

Suppose now that z_o is a point of linear density for E in both x and y-directions. We are going to prove that f is differentiable at z_o. By the definition of z_o there is a square $S_{\delta^*} = \{x+iy: \max(|x-x_o|, |y-y_o|) < \delta^* < \delta\}$ such that for any closed line segment I, $I \subset S_{\delta^*}$, containing z_o and parallel to either coordinate axis, we have

$$(3.5)\quad \|I \cap E\| \geq \|I\|/(1+\varepsilon).$$

We suppose next, in addition, that $z \in S_{\frac{1}{2}\delta^*} \setminus \{0\}$ and $\varepsilon < 1$. By (3.5), each of the line segments $(x+iy_o - \varepsilon(x-x_o); x+iy_o)$, $(x+iy_o; x+iy_o + \varepsilon(x-x_o))$, $(x_o+iy-\varepsilon i(y-y_o); x_o+iy)$, $(x_o+iy; x_o+iy+\varepsilon i(y-y_o))$ contains at least one point of E; let us denote those we have chosen by x_1+iy_o, x_2+iy_o, x_o+iy_1, x_o+iy_2, respectively. Let $U^* = \{x+iy: x_1 \leq x \leq x_2, y_1 \leq y \leq y_2\}$. Since f is open, it satisfies the maximum principle. Hence there is a $z^* \in \operatorname{fr} U^*$ such that $|f(s) - L(z)| \leq |f(z^*) - L(z)|$ for $s \in \operatorname{int} U^*$, where $L(z) = f(z_o) + f_z(z_o)(z-z_o) + f_{\bar z}(z_o)(\bar z - \bar z_o)$. In particular, we have

$$(3.6)\quad \begin{aligned} |f(z) - L(z)| &\leq |f(z^*) - L(z)| \leq |f(z^*) - L(z^*)| + |L(z^*) - L(z)| \\ &\leq |f(z^*) - L(z^*)| + 2\varepsilon\{|f_z(z_o)| + |f_{\bar z}(z_o)|\}|z - z_o|. \end{aligned}$$

By the definition of U^*, either $x^*+iy_o \in E$ or $x_o+iy^* \in E$. Suppose e.g. that $x^*+iy_o \in E$. By (3.3) and (3.4), we get

$$\begin{aligned} |f(z^*) - L(z^*)| &\leq |f(x^*+iy^*) - f(x^*+iy_o) - f_y(x^*+iy_o)(y^*-y_o)| \\ &\quad + |f(x^*+iy_o) - f(x_o+iy_o) - f_x(x_o+iy_o)(x^*-x_o)| \\ &\quad + |f_y(x^*+iy_o) - f_y(x_o+iy_o)|\,|y^*-y_o| \\ &\leq |f(x^*+iy_o, y^*-y_o)|\,|y^*-y_o| + |f(x_o+iy_o, x^*-x_o)|\,|x^*-x_o| \\ &\quad + |f_y(x^*+iy_o) - f_y(x_o+iy_o)|\,|y^*-y_o| \\ &\leq (1+2\varepsilon)\{|f(x^*+iy_o, y^*-y_o)| + |f(x_o+iy_o, x^*-x_o)| \\ &\quad + |f_y(x^*+iy_o) - f_y(x_o+iy_o)|\}|z-z_o|. \end{aligned}$$

Hence, by (3.4), we obtain

$$(3.7)\quad |f(z^*) - L(z^*)| < 3(1+2\varepsilon)\varepsilon|z-z_o|.$$

An analogous calculation leads to (3.7) if $x_o+iy^* \in E$. Relations (3.6) and (3.7) yield (1.1), so f is differentiable at z_o, as desired.

3. Distributional derivatives

Finally we note that almost all points of a measurable set in $\mathbb{C}$ are their points of linear density in both x and y-directions (cf. Saks [2], p. 298). Hence our assertion follows.

Suppose now that $f\colon D \longrightarrow D' \subset \mathbb{C}$, where D is a domain in $\mathbb{C}$, is an ACL sense-preserving homeomorphism and (2.2) holds. Let z_0 be a point of differentiability and U a closed square containing z_0 and with sides t. We have $|L[U]| = t^2 J(z_0)$, $\operatorname{dia} L[U] \le 2t\{|f_x(z_0)| + |f_y(z_0)|\}$, and $|f(z) - L(z)| \le |z - z_0|\varepsilon(t)$, $z \in U$, where $\varepsilon(t)$ is the maximum of $(1/|z - z_0|)|f(z) - L(z)|$ over all $|z - z_0| \in (0; 2t)$. Therefore

$$|f[u]| \le t^2 J(z_0) + 16\{|f_x(z_0)| + |f_y(z_0)|\}t^2\varepsilon(t) + 4\pi t^2\varepsilon^2(t),$$

$$|f[U]| \ge t^2 J(z_0) - 16\{|f_x(z_0)| + |f_y(z_0)|\}t^2\varepsilon(t) + 16t^2\varepsilon^2(t),$$

whence $(1/t)|f[U]| \longrightarrow J(z_0)$ as $t \longrightarrow 0+$. On the other hand, a theorem of Lebesgue (cf. e.g. Saks [2], pp. 115 and 119) states that a non-negative totally additive bounded set function $\tau\colon D \longrightarrow \mathbb{R}$ possesses a.e. a finite measurable derivative τ' and

$$(3.8)\qquad \tau(E) \ge \iint_E \tau' \,dxdy$$

for any Borel set $E \subset D$ with equality iff τ is locally absolutely continuous (concerning D in this theorem only $D \subset C$ is assumed). With $\tau(E) = |f[E]|$ we have $\tau'(z) = J(z)$ a.e. in D so that, by (3.8), J is locally L^1. Since, by (1.5) and (2.2), $(|f_z|^2 + |f_{\bar z}|)^2 \le Q\,J$ a.e., thus f_z and $f_{\bar z}$ are locally

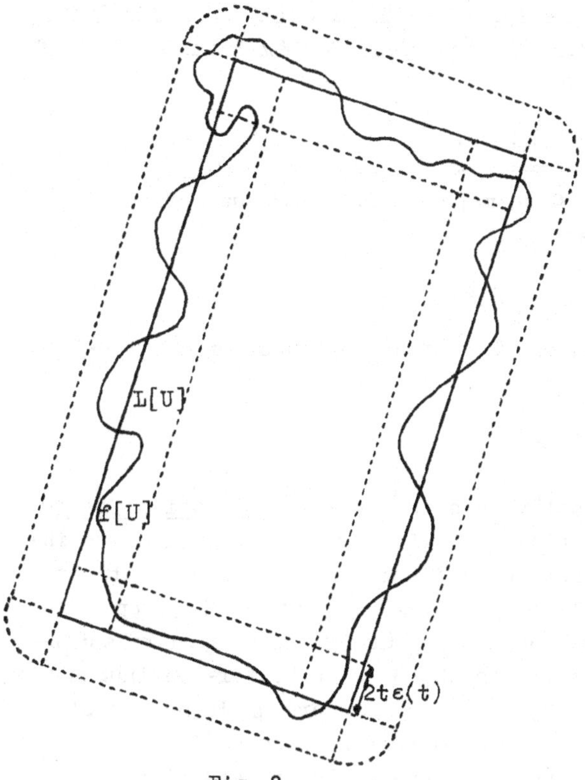

Fig. 2

L^2. Finally, let $h: D \to \mathbb{C}$ be a C^1-function with compact support. Since f is ACL, we have

$$\int_I (fh)_x dx = 0$$

for a.e. horizontal segment in D whose end points are in D spt h. By applying Fubini's theorem we conclude that

$$\iint_D f_x h\,dxdy = -\iint_D fh_x dxdy.$$

Similarly we have

$$\iint_D f_y h\,dxdy = -\iint_D fh_y dxdy,$$

whence (3.1) with $f'_z = f_z$ and $f'_{\bar{z}} = f_{\bar{z}}$ a.e. follows. Thus we conclude this section with the following result (cf. Bers [1]):

LEMMA 3. If $f: D \to D' \subset \mathbb{C}$, where D is a domain in $\mathbb{C}$, is an ACL sense-preserving homeomorphism and (2.2) holds, then f possesses distributional L^p-derivatives f'_z, $f'_{\bar{z}}$, where we may take $p = 2$. Moreover, $f'_z = f_z$ and $f'_{\bar{z}} = f_{\bar{z}}$ a.e.

4. The Beltrami differential equation

By Lemmas 1 and 3, any Q-qc mapping $f: D \to \mathbb{C}$ satisfies (2.2) and $f'_z = f_z$, $f'_{\bar{z}} = f_{\bar{z}}$ a.e. Hence

$$(4.1) \quad |f'_{\bar{z}}| \le [(Q-1)/(Q+1)]\,|f'_z| \quad \text{a.e.}$$

Under the complex dilatation of f we mean any function $\mu: D \to \mathbb{E}$ satisfying

$$(4.2) \quad w'_{\bar{z}} = \mu w'_z \quad \text{a.e.},$$

where $w = f(z)$; (4.2) is usually referred to as a Beltrami differential equation (cf. Beltrami [1]) and, given μ, it is natural to investigate the problem of existence and uniqueness of its solutions. For the history of this problem cf. Lehto and Virtanen [1], pp. 204-205, or [2], pp. 194-195, and Lehto [2]. In our discussion we essentially follow Vekua [2] in the case where μ is Hölder-continuous and locally L^p, and Bojarski [1, 2] in the case where μ is measurable with $\|\mu\|_\infty < 1$. In some stages we also follow Ahlfors [5].

In each case we replace (4.2) by the system of equations

(4.3) $w'_{\bar z} = g$ a.e.,

(4.4) $\mu w'_z = g$ a.e.

We have

LEMMA 4. *Let* D *be a domain in* $\mathbb{C}$ *and* $g: D \to \mathbb{C}$ *be locally* L^p. *Then* $w = f(z)$, $z \in D$, *satisfies* (4.3) *iff*

$$(4.5) \quad f = \Phi + Tg, \quad Tg(z) = -\frac{1}{\pi}\lim_{\varepsilon \to 0+} \iint_{D(z;\varepsilon)} \frac{g(\zeta)}{\zeta - z}\, d\xi\, d\eta,$$

where Φ *is holomorphic*, $\xi = \operatorname{re}\zeta$, $\eta = \operatorname{im}\zeta$, *and* $D(z;\varepsilon) = D \setminus \Delta^\varepsilon(z)$.

Proof. We begin with proving that

(4.6) $(Tg)'_{\bar z} = g$ a.e.,

i.e.

$$(4.7) \quad \iint_D gh\, dxdy = -\iint_D Tgh_{\bar z}\, dxdy$$

for all C^1-functions $h: D \to \mathbb{C}$ with compact support. To this end, given $s \in D$, consider a domain D_0 with $D_0 \supset \operatorname{spt} h$, $\operatorname{cl} D_0 \subset D$, and rectifiable ∂D_0 consisting of disjoint Jordan curves. By the second of Green's formulae (3.2) applied to $D^* = D_0(s;\varepsilon)$ and $h^*(z) = h(z)/(z-s)$, we have

$$\iint_{D_0(s;\varepsilon)} h_{\bar z}(z)\frac{dxdy}{z-s} + \frac{1}{2i}\int_{\partial\Delta^\varepsilon(s)} h(z)\frac{dz}{z-s} = \frac{1}{2i}\int_{\partial D_0} h(z)\frac{dz}{z-s},$$

whence, on letting $\varepsilon \to 0+$ and utilizing $h|\operatorname{fr} D_0 = 0$, we get $Th_{\bar z} = h$. Consequently, on applying the definition of T in the right-hand side of (4.7) and interchanging the order of integration, we obtain (4.7), as desired.

Suppose now that $w = f(z)$ satisfies (4.3). Hence, by (4.6),

(4.8) $(f - Tg)'_{\bar z} = 0$ a.e.

We claim that $\Phi^* = f - Tg$ is equal to a holomorphic function a.e. To this end it is clearly sufficient to prove that so it is inside of some $\Delta^r(z_0) \subset D$ and we may of course assume that $s_0 = 0$. Let

$$Z(z,s) = 2|z-s|^2 \log\frac{|r^2 - z\bar s|}{r|z-s|} - (r^2 - |z|^2)\left(1 - \frac{|s|^2}{r^2}\right);$$

$$z \in \Delta^r,\ s \in D,\ z \neq s.$$

It is easily verified that any $Z(\ ,s)$ satisfies $\nabla^2\nabla^2 Z = 0$ with the boundary conditions $Z = Z_x = Z_y = 0$, where all the functions are restricted to $\mathrm{fr}\,\Delta^r$. Hence it can be extended by the value 0 to a C^1-function $Z(\ ,s): D \to \mathbb{C}$ with compact support. Therefore, by (4.8),

$$\iint_{\Delta^r} \Phi^* Z_{\bar{z}}\,dxdy = \iint_D \Phi^* Z_{\bar{z}}\,dxdy = -\iint_D \Phi^{*\prime}_{\bar{z}} Z\,dxdy = 0.$$

Since

$$Z_{\bar{z}s\bar{s}}(z,s) = \frac{1}{\bar{s}-\bar{z}} + \frac{r^2 z - 2r^2 s + \bar{z}s^2}{(r^2 - \bar{z}s)^2} + \frac{z^2\bar{s}}{r^2(r^2 - z\bar{s})},$$

we obtain, after interchanging the order of $\frac{\partial^2}{\partial s\partial\bar{s}}$ and $\iint_{\Delta^r}$,

$$\overline{T\Phi^*(s)} = -\frac{1}{\pi}\iint_{\Delta^r} \Phi^*(z)\frac{r^2 z - 2r^2 s + \bar{z}s^2}{(r^2 - \bar{z}s)^2}\,dxdy - \frac{1}{\pi r^2}\iint_{\Delta^r} \Phi^*(z)\frac{z^2\bar{s}\,dxdy}{r^2 - z\bar{s}}.$$

Consequently, by (1.2) and (4.6), we conclude that

$$\Phi^*(s) = -\frac{1}{\pi}\frac{\partial}{\partial s}\iint_{\Delta^r} \Phi^*(z)\frac{r^2 z - 2r^2 s + \bar{z}s^2}{(r^2 - \bar{z}s)^2}\,dxdy \quad \text{a.e. in } \Delta^r,$$

i.e. $\Phi^*|\Delta^r$ is equal to a holomorphic function a.e., as desired.

The converse result is a straightforward consequence of (4.6).

Next it is natural to calculate f'_z, where f is given by (4.5). Since

$$\frac{f(s) - f(s_0)}{s - s_0} = \frac{\Phi(s) - \Phi(s_0)}{s - s_0} - \frac{1}{\pi}\lim_{\varepsilon\to 0+}\iint_{D(z;\varepsilon)} \frac{g(z)}{(z - s)(z - s_0)}\,dxdy,$$

it is natural to suggest that

$$\text{(4.9)}\quad f'_z = \Phi' + Sg \ \text{ a.e.,}\quad Sg(z) = -\frac{1}{\pi}\lim_{\varepsilon\to 0+}\iint_{D(z;\varepsilon)} \frac{g(\zeta)}{(\zeta - z)^2}\,d\xi d\eta,$$

where, as before, $D(z;\ ) = D \setminus \Delta^\varepsilon(z)$. For the sake of simplicity, following Ahlfors [5], we suppose first that g is C^1 with compact support and extend it by the value 0 to $g: \Delta^r \to \mathbb{C}$, $\Delta^r \supset D$. Since, by the first Green's formula (3.2) applied to $h^*(z) = 1/(z - s_0)$ and $D^* = \mathrm{int}\,\Delta^r(s_0;\varepsilon)$ such that $\Delta^\varepsilon(s_0) \subset \Delta^r$, we have

$$\iint_{\Delta^r(s_0;\varepsilon)} \frac{dxdy}{(z - s_0)^2} = -\iint_{\Delta^r(s_0;\varepsilon)} \frac{\partial}{\partial z}\frac{1}{z - s_0}\,dxdy$$

$$= \frac{1}{2i}\int_{\partial\Lambda^r}\frac{d\bar z}{z-s_0} - \frac{1}{2i}\int_{\partial\Lambda^\varepsilon(s_0)}\frac{d\bar z}{z-s_0} = 0,$$

then

$$\lim_{\varepsilon\to 0+}\iint_{\Lambda^r(z;\varepsilon)}\frac{g(z)}{(z-s_0)^2} = \iint_{\Lambda^r}\frac{g(z)-g(s_0)}{(z-s_0)^2}\,dxdy$$

provided that the last integral converges, but this is a consequence of C^1-continuity of g which implies $|g(z)-g(s_0)| \le M|z-s_0|$, $M<+\infty$, in Λ^r.

Observe now that, by the second Green's formula (3.2),

$$\iint_{\Lambda^r}\frac{g(s_0)}{(z-s)(z-s_0)}\,dxdy = \frac{g(s_0)}{s-s_0}\iint_{\Lambda^r}\left(\frac{1}{z-s}-\frac{1}{z-s_0}\right)dxdy$$

$$= \frac{g(s_0)}{s-s_0}\iint_{\Lambda^r}\left(\frac{\partial}{\partial\bar s}\frac{\bar s}{z-s}-\frac{\partial}{\partial\bar s_0}\frac{\bar s_0}{z-s_0}\right)dxdy = \frac{1}{2i}\frac{g(s_0)}{s-s_0}\int_{\partial\Lambda^r}\left(\frac{\bar s}{z-s}-\frac{\bar s_0}{z-s_0}\right)dz$$

$$= \pi g(s_0)(\bar s-\bar s_0)/(s-s_0),$$

where $s\in D$, so in order to conclude (4.9) it suffices to prove that

$$\lim_{s\to s_0}\iint_{\Lambda^r}\frac{g(z)-g(s_0)}{(z-s)(z-s_0)}\,dxdy = \iint_{\Lambda^r}\frac{g(z)-g(s_0)}{(z-s_0)^2}\,dxdy.$$

To this end, following Lehto and Virtanen [1] or [2], we consider the difference of the above integrals and estimate it by

$$|s-s_0|\iint_{\Lambda^r}\frac{|g(z)-g(s_0)|}{|z-s_0|^2|z-s|}\,dxdy \le M|s-s_0|\iint_{\Lambda^r}\frac{dxdy}{|z-s_0||z-s|}.$$

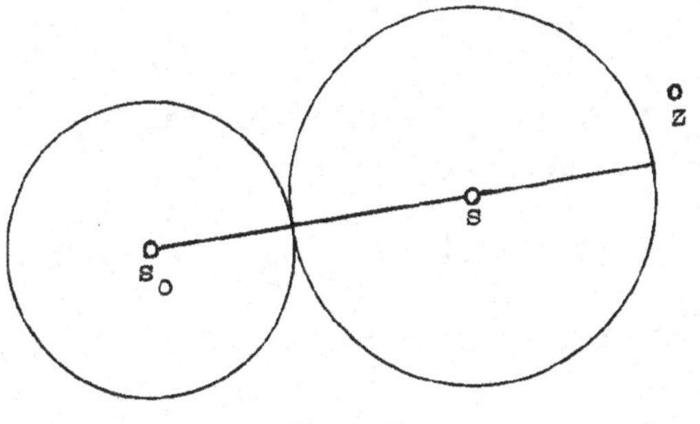

Fig. 3

For $|s-s_0|$ sufficiently small we have $\Lambda(s_0)\cup\Lambda(s)\subset\Lambda^r$, where $\Lambda(a) = \{z:\ |z-a| \le \frac{1}{2}|s-s_0|\}$, and also $|z-s_0||z-s| > \frac{1}{3}|z-s_0|^2$ whenever $z\in\Lambda^r\setminus\Lambda(s_0)\setminus\Lambda(s)$. Consequently, under the notation $\Lambda(s,s_0) = \{z:\ \frac{1}{2}|s-s_0| \le |z-s_0| \le 2r\}$, the difference in question is estimated by

$$2M\iint_{\Delta(s_0)} \frac{dxdy}{|z-s_0|} + 2M\iint_{\Delta(s)} \frac{dxdy}{|z-s|} + 3M|s-s_0| \iint_{\Delta(s,s_0)} \frac{dxdy}{|z-s_0|^2}$$

$$= 4\pi M|s-s_0|[1+\tfrac{3}{2}\log(4r/|s-s_0|)] \longrightarrow 0 \text{ as } s \longrightarrow s_0,$$

as desired. Thus we conclude:

LEMMA 5. *Let* D *be a domain in* $\mathbb{C}$ *and* $g: D \longrightarrow \mathbb{C}$ *a* C^1*-function with compact support. Then* (4.5) *implies* (4.9) *everywhere in* D, S *being well-defined,* Φ *being holomorphic, and* $\xi = \operatorname{re}\zeta$, $\eta = \operatorname{im}\zeta$.

Actually Lemma 5 remains valid for g being L^p, $p > 1$. For such an extension we need some results due to M. Riesz [1, 2] and Calderón and Zygmund [2]; cf. also Zygmund [1] (pp. 147-149 and 192-200) and [2] (vol. I, pp. 253-262, and vol. II, pp. 93-96). We begin with two lemmas of M. Riesz [2, 1].

5. Two lemmas of M. Riesz

LEMMA 6. *For any* C^1*-function* $g: \mathbb{R} \longrightarrow \mathbb{R}$ *with compact support, we have* $\|Hg\|_p \leq A_p\|g\|_p$, $1 < p < +\infty$, A_p *being finite and depending on* p *only with* $A_2 = 1$, *where*

$$Hg(x) = \frac{1}{\pi} \lim_{\varepsilon \to 0+} \left\{ \int_{-\infty}^{x-\varepsilon} + \int_{x+\varepsilon}^{+\infty} \right\} \frac{g(\xi)}{\xi - x} d\xi.$$

Proof (the idea of this proof belongs to Stein [1]). Suppose first that $p > 2$. If we extend Hg by the same formula to the upper half-plane, we can write

$$Hg(z) = \frac{1}{\pi}\int_0^{+\infty} \frac{1}{1+(y/s)^2} \frac{g(x+s)-g(x-s)}{s} ds + \frac{1}{\pi} iy \int_{-\infty}^{+\infty} \frac{g(s)}{(s-x)^2+y^2} ds$$

$$= u(z) + iv(z).$$

Hence $u(z) \longrightarrow Hg(x)$ and, by Poisson's formula for the upper half-plane, $v(z) \longrightarrow g(z)$ as $y \longrightarrow 0+$. Consider now the integral

$$I_\varepsilon = \int_{\partial D_\varepsilon} - \{|Hg|^p - (1-1/p)^{-1}|u|^p\}_y dx + \{|Hg|^p - (1-1/p)^{-1}|u|^p\}_x dy,$$

where $D_\varepsilon = \{z : y \geq \varepsilon\} \cap \Delta^{1/\varepsilon}$. By Green's formula, we have

$$I_\varepsilon = \iint_{D_\varepsilon} \nabla^2 \{|Hg|^p - (1-1/p)^{-1}|u|^p\} dxdy.$$

Since $\nabla^2|u|^p = p(p-1)|u|^{p-2}\nabla.\nabla u$, $\nabla^2|Hg|^p = p^2|Hg|^{p-2}\nabla.\nabla u$, and $p \geq 2$, then $\nabla^2\{|Hg|^p - (1-1/p)^{-1}|u|^p\} = p^2(|Hg|^{p-2} - |u|^{p-2})\nabla\cdot\nabla u \geq 0$ and, consequently, $I_\varepsilon \geq 0$. On the other hand, it can directly be verified that

$$I_\varepsilon + \int_{-1/\varepsilon + i\varepsilon}^{1/\varepsilon + i\varepsilon} \{|Hg|^p - (1-1/p)^{-1}|u|^p\}_y \, dz \longrightarrow 0 \text{ as } \varepsilon \longrightarrow 0+,$$

so, after interchanging the order of differentiation and integration, we obtain

$$(\partial/\partial y)\int_{-\infty}^{+\infty} \{|Hg|^p - (1-1/p)^{-1}|u|^p\}\, dx \leq 0.$$

Next we check that the above integral tends to 0 as $y \longrightarrow +\infty$. Hence, for every $y > 0$, we get

$$\int_{-\infty}^{+\infty} |Hg|^p dx \geq (1-1/p)^{-1}\int_{-\infty}^{+\infty} |u(z)|^p dx,$$

i.e. $\|u^2\|_{\frac{1}{2}p} \leq (1-1/p)^{2/p}\|(Hg)^2\|_{\frac{1}{2}p}$ and, by Schwarz's inequality,

$$\|u^2\|_{\frac{1}{2}p} = (1-1/p)^{2/p}(\|u^2\|_{\frac{1}{2}p} + \|v^2\|_{\frac{1}{2}p}).$$

Therefore $\|u^2\|_{\frac{1}{2}p} \leq [(1-1/p)^{-2/p} - 1]^{-1}\|v^2\|_{\frac{1}{2}p}$ and, after letting $y \longrightarrow 0+$, we obtain the desired conclusion with $A_p = [(1-1/p)^{-2/p} - 1]^{-\frac{1}{2}}$.

Let now $1 < p < 2$ and let p' be determined by $1/p + 1/p' = 1$. By Hölder's inequality,

$$\left|\int_{-\infty}^{+\infty} (Hg)h\, dx\right| \leq \|Hg\|_p \|h\|_{p'} \text{ for every } L^{p'}\text{-function } h: \mathbb{R} \longrightarrow \mathbb{R},$$

in particular for every C^1-function $h: \mathbb{R} \longrightarrow \mathbb{R}$ such that $\|h\|_{p'} = 1$, with equality for $|h| = c|Hg|^{p/p'}$, c being any positive constant. Hence

$$\|Hg\|_p = \sup\left|\int_{-\infty}^{+\infty} (Hg)h\, dx\right|,$$ where the supremum is taken over all C^1-functions h, described above.

Next we observe that

$$\int_{-\infty}^{+\infty} (Hg)h\, dx = \frac{1}{\pi}\int_{-\infty}^{+\infty} \lim_{\varepsilon \to 0+}\left\{\int_{-\infty}^{x-\varepsilon} + \int_{x+\varepsilon}^{+\infty}\right\} \frac{g(\xi)h(x)}{\xi - x}\, d\xi\, dx$$

$$= -\frac{1}{\pi}\int_{-\infty}^{+\infty} \lim_{\varepsilon \to 0+}\left\{\int_{-\infty}^{\xi-\varepsilon} + \int_{\xi+\varepsilon}^{+\infty}\right\} \frac{h(x)g(\xi)}{x - \xi}\, dx\, d\xi = -\int_{-\infty}^{+\infty} g\, H h\, d\xi$$

and that, by Hölder's inequality and our lemma for $p' \geq 2$, we have

$$\Big|\int_{-\infty}^{+\infty} g\,Hh\,d\xi\Big| \le \|Hh\|_{p'}\cdot\|g\|_{p} \le A_{p'}\cdot\|g\|_{p}.$$

Consequently we obtain the desired conclusion with $A_p = A_{p'}$.

LEMMA 7. Given two numbers p_k, $k=1,2$, $1<p_1<p_2<+\infty$, suppose that F is a linear operator which associates with every L^{p_k}-function $g\colon ₵ \to ₵$ another function of the same class and fulfills the conditions $\|F\|_{p_k} \le +\infty$, where

$$\|F\|_p = \sup_g(\|Fg\|_p/\|g\|_p).$$

Then for any p, $p_1 \le p \le p_2$, F associates with every L^p-function $g\colon$ $₵ \to ₵$ another function of the same class and fulfills the condition $\|F\|_p < +\infty$. Moreover,

$$(5.1)\quad \|F\|_p \le \|F\|_{q_1}^{1-t}\|F\|_{q_2}^{t},\quad t = \Big(\frac{1}{q_1}-\frac{1}{p}\Big)\Big/\Big(\frac{1}{q_1}-\frac{1}{q_2}\Big),\quad p_1 \le q_1 < p < q_2 \le p_2.$$

One can say that $\|F\|_p$ is multiplicatively convex w.r.t. $1/p$ or that $\log\|F\|_p$ is convex w.r.t. $1/p$ within $[p_1; p_2]$.

1st proof. For greater clarity the proof is divided into five steps.

Step A. Let ₵ denote the class of all (Lebesgue) measurable functions $g\colon ₵ \to ₵$ with compact support and finite number of values. We claim that ₵ is dense in every class of all L^p-functions from ₵ into itself, $1 \le p < +\infty$; the distance of g and g^* being defined as $\|g-g^*\|_p$.

Indeed, consider a function g^* of the class L^p in question and suppose first that g^* has a compact support. Choose a sequence of nets of squares covering $\operatorname{spt} g^*$ with diameters tending to 0 as the index $n \to +\infty$ and define $(g_n) \subset ₵$ by the formulae

$$g_n(s) = (1/|D|)\iint_D g^*\,dx\,dy,\quad s \in D,$$

for each open square D, $D \subset \operatorname{spt} g^*$ of the n-th net, while $g_n(s) = 0$ otherwise. By Hölder's inequality, we have $\|g_n\|_p \le \|g^*\|_p$. On the other hand, $g_n \to g^*$ a.e., so, if g^* is bounded, we have $\|g^* - g_n\|_p \to 0$. If g^* is unbounded we write $g^* = g_1^* + g_2^*$, where g_1^* is bounded and $\|g_2^*\|_p < \frac{1}{3}\varepsilon$, ε being a positive parameter. Since $g_n = g_{1,n} + g_{2,n}$, where $g_{k,n}$ are analogues of g_n, and

$$\|g^* - g_n\|_p \le \|g_1^* - g_{1,n}\|_p + \|g_2^*\|_p + \|g_{2,n}\|_p < \|g_1^* - g_{1,n}\|_p + \tfrac{2}{3}\varepsilon < \varepsilon$$

for n sufficiently large, then also in this case $\|g^* - g_n\|_p \to 0$. Finally, suppose that $\operatorname{spt} g^*$ is not compact, g^* being bounded or not. We again write $g^* = g_3^* + g_4^*$, where g_3^* has compact support, $(g^* - g_3^*)|$ $|\operatorname{spt} g_3^* = 0$, and $\|g_4^*\|_p < \frac{1}{2}\varepsilon$. Take a function $g_0 \in ¢$, $\operatorname{spt} g_0 \subset \operatorname{spt} g_3^*$, so that $\|g_3^* - g_0\|_p < \frac{1}{2}\varepsilon$. Then $\|g^* - g_0\|_p \leq \|g_3^* - g_0\|_p + \|g_4^*\|_p < \varepsilon$, and this shows the desired density of $¢$.

Step B. Now <u>we claim that it is sufficient to consider functions of</u> $¢$. In fact, for g^* as before and $(g_n) \in ¢$ such that $\|g^* - g_n\|_p \to 0$, we have $\|g_m - g_n\|_p \to 0$ as $m, n \to +\infty$. Thus, if we prove that $\|Fg_n\|_p \leq \|F\|_p \|g_n\|_p$ for any n, we get $\|Fg_m - Fg_n\|_p \to 0$. Since L^p is complete, there exists an L^p-function, which we may denote by $Fg^*: ¢ \to ¢$, such that $\|Fg^* - Fg_n\|_p \to 0$. This function is, of course, defined outside a set of measure zero and independent of the choice of (g_n). If $\|Fg\|_p \leq \|F\|_p \|g\|_p$ for all $g = g_n$, it also holds for g^*.

Step C. Consider then a function $g \in ¢$ and the corresponding sets of constancy of g. If χ_k, $k = 1, \dots, n$, denote the characteristic functions of the sets in question, we can write $g = z_1\chi_1 + \dots + z_n\chi_n$, where z_k are (complex) constants. Also $Fg = z_1F\chi_1 + \dots + z_nF\chi_n$. Since χ_k are L^{p_1} and L^{p_2}-functions, $F\chi_k$ are also such, by hypothesis, and from $p_1 \leq p \leq p_2$ it follows, by Hölder's inequality, that $F\chi_k$ are L^p-functions. <u>We claim that, given</u> $\varepsilon > 0$, <u>there are</u> $\tilde{g}_k$, $k = 1, \dots, n$, <u>such that</u> $\|F\chi_k - \tilde{g}_k\|_p < \varepsilon$ <u>and</u>

$$(5.2) \qquad \|\tilde{g}\|_q \leq \|Fg\|_q,$$

<u>where</u> $\tilde{g} = z_1\tilde{g}_1 + \dots + z_n\tilde{g}_n$, $1 \leq q < +\infty$, <u>and</u> (z_k) <u>is a system of</u> (complex) <u>constants</u>.

Indeed, suppose first that g is bounded and has compact support. Choose a subdivision of the system of sets in question by means of a net of squares and define $(\tilde{g}_1, \dots, \tilde{g}_n)$ by the formulae

$$\tilde{g}_k(s) = (1/|D|) \iint_D F\chi_k \, dxdy, \quad s \in D,$$

for each open square D, $D \subset \operatorname{spt} F\chi_k$ of our subdivision, while $\tilde{g}_k(s) = 0$ otherwise. If this subdivision is sufficiently dense, we get $\|F\chi_k - \tilde{g}_k\|_p < \varepsilon$. Furthermore, by Hölder's inequality, we obtain (5.2) for $1 \leq q < +\infty$ and we notice that (5.2) holds trivially for $q = +\infty$. In the case where neither boundedness of g nor compactness of $\operatorname{spt} g$ is assumed, we write, as in the last stage of Step A, $F\chi_k = \hat{g}_k + (F\chi_k - \tilde{g}_k)$, where each $\hat{g}_k$ has compact support and $(F\chi_k - \hat{g}_k) | \operatorname{spt} \hat{g}_k = 0$. Let $\hat{g} =$

$z_1\hat{g}_1 + z_2\hat{g}_2 + \ldots$ and let

$$g_k^o(s) = (1/|D|)\iint_D \hat{g}_k \,dxdy, \quad s \in D,$$

for each open square D, $D \subset \operatorname{spt} \hat{g}_k$ of the subdivision in question, while $g_k^o(s) = 0$ otherwise. If $\tilde{g} = z_1 g_1^o + z_2 g_2^o + \ldots$ then, by Hölder's inequality, we have again $\|\tilde{g}\|_q \le \|\hat{g}\|_q \le \|Fg\|_q$, as desired.

Step D. Let

$$z^* = \sup_{z_1,\ldots,z_n} [(|z_1| + \ldots + |z_n|)/(|D_1||z_1|^p + \ldots + |D_n||z_n|^p)^{1/p}],$$

where $D_1,\ldots,D_n$ are the squares in question. It is clear that the above supremum is attained. Given $\varepsilon > 0$, let $\eta = z^*\varepsilon$. Since, by Minkowski's inequality, $|\|Fg\|_p - \|\tilde{g}\|_p| < \varepsilon(|z_1| + \ldots + |z_n|)$, we see that

$$(5.3) \quad \|Fg\|_p/\|g\|_p \le \eta + \|\tilde{g}\|_p/(|D_1||z_1|^p + \ldots + |D_n||z_n|^p)^{1/p}.$$

Consider now a system of (complex) numbers $a_{j,k}$; $j = 1,\ldots,m$; $k = 1,\ldots,n$, and the linear forms $Z_j = a_{j,1}z_1 + \ldots + a_{j,n}z_n$. Let further

$$(5.4) \quad M_p = \sup_{z_1,\ldots,z_n} [(a_1|Z_1|^p + \ldots + a_m|Z_m|^p)^{1/p}/(|D_1||z_1|^p + \ldots + |D_n||z_n|^p)^{1/p}],$$

$a_1,\ldots,a_m$ being arbitrary (fixed) positive numbers (in our case we shall choose a_j so that $\|\tilde{g}\|_p = (a_1|Z_1|^p + \ldots + a_m|Z_m|^p)^{1/p})$. It is clear that the supremum in (5.4) is attained for $(z_1^*,\ldots,z_n^*)$, say. We claim that

$$(5.5) \quad M_p \le M_{q_1}^{1-t} M_{q_2}^t, \quad t = \left(\frac{1}{q_1} - \frac{1}{p}\right)/\left(\frac{1}{q_1} - \frac{1}{q_2}\right), \quad 1 \le q_1 < q_2 \le +\infty,$$

i.e. that M_p is multiplicatively convex w.r.t. $1/p$ within $[1; +\infty]$.

In fact, consider the function

$$f(\varepsilon) = (\Sigma a_j|Z_j^* + \varepsilon Z_j^o|^p)^{1/p}/(\Sigma |D_k||z_k^* + \varepsilon z_k^o|^p)^{1/p}, \quad \varepsilon \in \mathbb{R},$$

where $Z_j^* = a_{j,1}z_1^* + \ldots + a_{j,n}z_n^*$, $Z_j^o = a_{j,1}z_1^o + \ldots + a_{j,n}z_n^o$, $z_k^o \in \mathbb{C}$. It is differentiable and attains its maximum for $\varepsilon = 0$, whence $f'(0) = 0$. Direct calculation gives

$$\Sigma a_j|Z_j^*|^p/\Sigma |D_k||z_k^*|^p = \operatorname{re} \Sigma a_j|Z_j^*|^{p-2}\bar{Z}_j^* Z_j^o/\operatorname{re} \Sigma |D_k||z_k^*|^{p-2}\bar{z}_k^* z_k^o$$

and this, with the particular choice $z_k^o = \tilde{z}_k^o \equiv z_k^* |z_k^*|^{\tau-1}$, $\tau > 1$, becomes

(5.6) $\Sigma a_j |Z_j^*|^p / \Sigma |D_k| |z_k^*|^p = \mathrm{re}\, \Sigma a_j |Z_j^*|^{p-2} \bar{Z}_j^* \tilde{Z}_j^o / \mathrm{re}\, \Sigma |D_k| |z_k^*|^{p+\tau-1}$,

where $\tilde{Z}_j^o = a_{j,1}\tilde{z}_1^o + \ldots + a_{j,n}\tilde{z}_n^o$. Now we write

$$\Sigma |D_k| |z_k^*|^{p+\tau-1} = (\Sigma |D_k| |z_k^*|^{p+\tau-1})^{1-\theta} (\Sigma |D_k| |z_k^*|^{p+\tau-1})^{\theta}$$

$$= (\Sigma |D_k| |z_k^*|^{p+\tau-1})^{1-\theta} (\Sigma |D_k| |\tilde{z}_k^o|^{(p+\tau-1)/\tau})^{\theta}, \quad 0 < \theta < 1,$$

and, by Hölder's inequality with exponents $1/(1-\theta)$ and $1/\theta$,

$$\Sigma a_j |Z_j^*|^{p-1} |\tilde{Z}_j^o| \le (\Sigma a_j |Z_j^*|^{(p-1)/(1-\theta)})^{1-\theta} (\Sigma a_j |\tilde{Z}_j^o|^{1/\theta})^{\theta}.$$

Therefore, with the particular choice $\theta = \tau/(p+\tau-1)$, (5.6) yields

$$\frac{\Sigma a_j |Z_j^*|^p}{\Sigma |D_k| |z_k^*|^p} \le \frac{(\Sigma a_j |Z_j^*|^{p+\tau-1})^{\frac{p-1}{p+\tau-1}} (\Sigma a_j |\tilde{Z}_j^o|^{\frac{p+\tau-1}{\tau}})^{\frac{\tau}{p+\tau-1}}}{(\Sigma |D_k| |z_k^*|^{p+\tau-1})^{\frac{p-1}{p+\tau-1}} (\Sigma |D_k| |\tilde{z}_k^o|^{\frac{p+\tau-1}{\tau}})^{\frac{\tau}{p+\tau-1}}},$$

whence we obtain (5.5) with $q_1 = p+\tau-1$, $q_2 = (p+\tau-1)/\tau$ and $t = 1/p$. For τ sufficiently near 1 we may then have q_1 and q_2 arbitrarily near p.

Suppose now, contrary to (5.5), that

(5.7) $M_p > M_{q_1^*}^{1-t(p)} M_{q_2^*}^{t(p)}$, $t(p) = \left(\frac{1}{q_1^*} - \frac{1}{p}\right) / \left(\frac{1}{q_1^*} - \frac{1}{q_2^*}\right)$,

for some q_1^*, q_2^*. Since $(\Sigma a_j |Z_j|^p)^{1/p} / (\Sigma |D_k| |z_k|^p)^{1/p}$ is continuous w.r.t. p in $(1; +\infty)$ and all z_j in $\mathbb{C}$, uniformly continuous on $E = \{(z_1, \ldots, z_n) : |z_1|^2 + \ldots + |z_n|^2 = 1\}$, and its maximum is attained on E, then M_p is continuous w.r.t. p in $(1; +\infty)$. Hence we may assume, subdividing – if necessary – $[q_1^*, q_2^*]$, that (5.7) holds for all p (with the same q_1^*, q_2^*) and that it remains valid for q_1^*, q_2^* replaced with any q_1, q_2 such that $q_1^* \le q_1 \le q \le q_2 \le q_2^*$ (and for all p), where q is the largest value of p, at which the ratio of the left- and right-hand sides in (5.7) attains its maximum. This, however, contradicts (5.5) with $p = q$, $q_1 = p+\tau-1$, $q_2 = (p+\tau-1)/\tau$, and, consequently, M_p is multiplicatively convex w.r.t. $1/p$ within $[1; +\infty]$.

S t e p E. Applying (5.5) to our case, we infer from (5.3) the estimate

$$\|Fg\|_p/\|g\|_p \le \eta + \sup_{z_1,\dots,z_n} [(\|\tilde{g}\|_{p_1}/\|g\|_{p_1})^{1-t}(\|\tilde{g}\|_{p_2}/\|g\|_{p_2})^{t}]$$

and this, by (5.2) and the definition of $\|F\|_p$, yields, after letting $\varepsilon \to 0+$, the estimate (5.1) with $q_1 = p_1$ and $q_2 = p_2$:

$$(5.8)\quad \|F\|_p \le \|F\|_{p_1}^{1-t}\|F\|_{p_2}^{t}, \quad t = \left(\frac{1}{p_1} - \frac{1}{p}\right)/\left(\frac{1}{p_1} - \frac{1}{p_2}\right).$$

Consequently, by Step B, F is a bounded linear operator from L^p into itself. Furthermore, applying (5.8) to $p = q_1$, $p_1 \le q_1 < p_2$, and then to $p = q_2$, $q_1 < q_2 \le p_2$, we conclude that the hypotheses of Lemma 7 are fulfilled with p_1 and p_2 replaced with q_1 and q_2, respectively, so (5.8) is to be replaced with (5.1) and this completes the proof.

2nd proof (Calderón and Zygmund [1]). Here Steps A and B remain unchanged, but then we utilize complex-analytic methods.

S t e p C′. Consider a function $g \in \mathbb{C}$ and let p' be determined by $1/p + 1/p' = 1$. By Hölder's inequality

$$\left|\iint_{\mathbb{C}} (Fg)h\,dxdy\right| \le \|Fg\|_p\|h\|_{p'}$$

for every $L^{p'}$-function $h: \mathbb{C} \to \mathbb{C}$, in particular for every $h \in \mathbb{C}$, such that $\|h\|_{p'} = 1$, with equality for $|h| = c|Fg|^{p/p'}$, c being any positive constant. Hence

$$\|Fg\|_p = \sup\left|\iint_{\mathbb{C}} (Fg)h\,dxdy\right|,$$

where the supremum is taken over all $h \in \mathbb{C}$ with $\|h\|_{p'} = 1$. We may suppose that $\|g\|_p = 1$. Fixing g and h, consider the integral

$$(5.9)\quad \mathfrak{I}(s) = \iint_{\mathbb{C}} [Fg^*(s)]h^*(s)\,dxdy, \quad 0 \le \operatorname{re} s \le 1,$$

where

$$g^*(s)(z) = |g(z)|^{p^*(s)-1}g(z), \quad p^*(s) = (p/p_1)(1-s) + (p/p_2)s, \quad \text{whenever } g(z) \ne 0,$$

$$h^*(s)(z) = |h(z)|^{p'^*(s)-1}h(z), \quad p'^*(s) = (p'/p_1')(1-s) + (p'/p_2')s, \quad \text{whenever } h(z) \ne 0,$$

while $g^*(s)(z) = 0$ whenever $g(z) = 0$ and $h^*(s)(z) = 0$ whenever $h(z) = 0$. <u>We claim that</u>

$$(5.10)\quad |\mathfrak{I}(s)| \le \|F\|_{p_1} \text{ for } \operatorname{re} s = 0, \quad |\mathfrak{I}(s)| \le \|F\|_{p_2} \text{ for } \operatorname{re} s = 1.$$

Indeed, consider the sets, where the constant values g_j, $j=1,\dots,m$, (say) of g and h_k, $k=1,\dots,n$, (say) of h are taken, and denote by χ_j and χ_k^*, respectively, the corresponding characteristic functions. Hence (5.9) becomes

$$\mathfrak{I}(s)=\sum_{j=1}^{m}\sum_{k=1}^{n}|g_j|^{p^*(s)-1}g_j|h_k|^{p'^*(s)-1}h_k\iint_{\mathbb{C}}(F\chi_j)\,\chi_k^*\,dxdy$$

$$=\sum_{j=1}^{mn}a_j\exp(\lambda_j s),\quad 0\le \mathrm{re}\, s\le 1;\quad a_j,\lambda_j\text{ - constants,}$$

so $\mathfrak{I}|\{s:\ \mathrm{re}\, s=s_0\}$ is bounded for every s_0. Applying Hölder's inequality to (5.9), we get, by hypotheses,

(5.11) $\quad |\mathfrak{I}(s)|\le \|Fg^*(s)\|_{p_k}\|h^*(s)\|_{p_k'}\le \|F\|_{p_k}\|g^*(s)\|_{p_k}\|h^*(s)\|_{p_k'},\quad k=1,2,$

where p_k' are determined by $1/p_k+1/p_k'=1$. If, in particular, $s_0=0$, then

$$\|g^*(s)\|_{p_1}=\|\,|g|^{p/p_1}\|_{p_1}=\|g\|_p^{p/p_1}=1,$$

$$\|h^*(s)\|_{p_1'}=\|\,|h|^{p'/p_1'}\|_{p_1'}=\|h\|_{p'}^{p'/p_1'}=1$$

and, similarly, for $s_0=1$, we obtain $\|g^*(s)\|_{p_2}=\|h^*(s)\|_{p_2'}=1$. Consequently (5.11) yields (5.10), as desired.

S t e p D′. <u>We now claim that</u> (5.8) <u>holds</u>. To this end we will prove that

(5.12) $\quad |\mathfrak{I}^*(s)|\le 1$ <u>for</u> $0\le \mathrm{re}\, s\le 1$, <u>where</u> $\mathfrak{I}^*(s)=\mathfrak{I}(s)/\|F\|_{p_1}^{1-s}\|F\|_{p_2}^{s}$,

what for $s=t$ gives the desired estimate.

We have already proved (5.12) for $\mathrm{re}\, s=0,1$, and have stated that $\mathfrak{I}|\{s:\ 0<\mathrm{re}\, s<1\}$ is bounded and holomorphic; hence so is $\mathfrak{I}^*|\{s:\ 0<\mathrm{re}\, s<1\}$. Therefore, if $\mathfrak{I}^*(s)\to 0$ as $\mathrm{im}\, s\to+\infty$ and $\mathrm{im}\, s\to-\infty$, (5.12) follows from the classical maximum principle applied to $\mathfrak{I}^*|\{s:\ 0\le\mathrm{re}\, s\le 1,\ -\eta\le\mathrm{im}\, s\le\eta\}$ for $\eta\to+\infty$. If $\mathfrak{I}^*(s)$ does not tend to 0 as $\mathrm{im}\, s\to+\infty$ or as $\mathrm{im}\, s\to-\infty$, we consider the sequence

$$\mathfrak{I}_n^*(s)=\mathfrak{I}^*(s)\exp(\tfrac{1}{n}s^2)=\mathfrak{I}^*(s)\exp(\tfrac{1}{n}\mathrm{re}^2 s-\tfrac{1}{n}\mathrm{im}^2 s)\exp(\tfrac{2}{n}i\,\mathrm{re}\, s\,\mathrm{im}\, s).$$

Each $\mathfrak{I}_n^*(s)\to 0$ as $\mathrm{im}\, s\to+\infty$ and $\mathrm{im}\, s\to-\infty$; and $|\mathfrak{I}_n^*(s)|\le e^{1/n}$, so, by the classical maximum principle applied to $\mathfrak{I}_n^*|\{s:\ 0\le\mathrm{re}\, s\le 1,\ -\eta\le\mathrm{im}\, s\le\eta\}$ for $\eta\to+\infty$, we infer that $|\mathfrak{I}_n^*(s)|\le\exp(1/n)$ for $0\le\mathrm{re}\, s\le 1$, and this, on letting $n \to +\infty$, concludes the proof of (5.12).

Step *E'*. From (5.8) we infer, as in Step E of the 1st proof, that F is a bounded linear operator from L^p into itself and that (5.1) follows also in the general case.

It is interesting to remark that the above proofs remain almost unchanged if F sends L^{p_k}-functions to L^{q_k}-functions, $p_k \geq q_k$, and the assertion is claimed to hold within $[(p_1,q_1);(p_2,q_2)]$. The 2nd proof remains also valid for $p_k < q_k$. The first complex-analytical proof of the result in question is due to Thorin [1], who also observed that it is restricted neither to $p_k \geq q_k$ nor to linear operators. Actually, Thorin proved the following result:

(i) *Let* f *be an entire function of* n *complex variables and* D *a bounded domain in* $\mathbb{R}_+^n$, $\mathbb{R}_+$ *denoting the right half-axis including the origin. Let further* $M(a_1,\ldots,a_n)$ *denote the least upper bound of* $|f| \mid \{(z_1,\ldots,z_n): |z_k| = v_k^{a_k},\ k=1,\ldots,n,\ (v_1,\ldots,v_n) \in D\}$. *Then* M *is multiplicatively convex w.r.t.* $(a_1,\ldots,a_n)$ *within* $\mathbb{R}_+^n$.

(ii) *If* $D \subset \{(v_1,\ldots,v_n): a \leq v_k \leq 1/a,\ k=1,\ldots,n\}$ *for some* $a>0$, *then* M *is multiplicatively convex w.r.t.* $(a_1,\ldots,a_n)$ *within* $\mathbb{R}^n$.

The original proof of Thorin was next simplified by Thorin [2], Tamarkin and Zygmund [1], and Calderón and Zygmund [1].

6. The Calderón-Zygmund inequality

We are now able to prove the extension of Lemma 5, announced at the end of Section 4. This extension is a particular case of a famous, deep, and important result of Calderón and Zygmund [2]. In the proof we follow the method of Vekua [2], based on that of Calderón and Zygmund [3], and then simplified by Ahlfors [5].

LEMMA 8. *Let* D *be a domain in* $\mathbb{C}$ *and* $g: D \to \mathbb{C}$ *be locally* L^p, $p>1$. *Then* (4.5) *implies* (4.9), S *being well-defined, where*

$$(6.1)\qquad \|S\|_p < +\infty,\quad 1<p<+\infty;\qquad \|S\|_p \to 1 \text{ as } p \to 2,$$

Φ *being holomorphic, and* $\xi = \operatorname{re}\zeta$, $\eta = \operatorname{im}\zeta$.

The inequality in (6.1) is called the *Calderón-Zygmund inequality*.

Proof. For greater clarity the proof is divided into six steps.

Step *A*. First of all we apply the fact that the class of all C^2-functions $g: D \to \mathbb{C}$ with $\operatorname{spt} g \subset D$ is dense in every class of all

6. The Calderón-Zygmund inequality

L^p-functions from D into $\mathbb{C}$, $1<p<+\infty$, the distance of g and g^* being defined as $\|g-g^*\|_p$ (cf. e.g. Lehto and Virtanen [1], p. 149, or [2], p. 142). Arguing exactly in the same way as in Step B of the 1st proof of Lemma 7 we conclude that <u>it is sufficient to consider functions of this subclass</u>.

S t e p B. Consider then a C^2-function $g: D \to \mathbb{C}$ with $\operatorname{spt} g \subset D$ and put

$$S^*g(z) = \frac{1}{2\pi} \lim_{\varepsilon \to 0+} \iint_{D(z;\varepsilon)} \frac{g(\zeta)}{(\zeta - z)|\zeta - z|} \, d\xi \, d\eta, \quad D(z;\varepsilon) = D \setminus \Delta^{\varepsilon}(z).$$

<u>We claim that</u>

$$(6.2) \qquad \|S^*g\|_p \leq \tfrac{1}{2}\pi A_p \|g\|_p,$$

<u>where</u> A_p <u>is the same as in Lemma</u> 6.

In fact, on putting

$$(6.3) \qquad g(z) = 0 \quad \text{for } z \,\bar{\in}\, D$$

and changing the variables of integration according to the formulae $\zeta = z + re^{i(t-\pi)}$, $-\pi < \arg\zeta \leq 0$, and $\zeta = z + re^{it}$, $0 < \arg\zeta \leq \pi$, we get

$$S^*g(z) = \frac{1}{2\pi} \lim_{\varepsilon \to 0+} \int_0^{\pi} \Big[\int_{\varepsilon}^{+\infty} g(z + re^{it}) \frac{dr}{r} - \int_{\varepsilon}^{+\infty} g(z - re^{it}) \frac{dr}{r} \Big] e^{-it} dt.$$

The subsequent change $r = \tilde{r} - \operatorname{re}(ze^{-it})$ in the first inner integral and $r = \operatorname{re}(ze^{-it}) - \tilde{r}$ in the second one yields

$$S^*g(z) = \frac{1}{2\pi} \lim_{\varepsilon \to 0+} \int_0^{\pi} \Big\{ \int_{-\infty}^{\operatorname{re}(ze^{-it})-\varepsilon} + \int_{\operatorname{re}(ze^{-it})+\varepsilon}^{+\infty} \Big\} \frac{1}{\tilde{r} - \operatorname{re}(ze^{-it})}$$

$$\times g(z + \tilde{r}e^{it} - \operatorname{re}(ze^{-it})e^{it}) \, d\tilde{r} \, dt,$$

i.e., after changing the order of $\int_0^{\pi}$ and $\lim_{\varepsilon \to 0+}$,

$$(6.4) \qquad S^*g(z) = \tfrac{1}{2} \int_0^{\pi} H\tilde{g}(\operatorname{re}(ze^{-it})) \, dt, \quad \tilde{g}(\tilde{r}) = g(z + \tilde{r}e^{it} - \operatorname{re}(ze^{-it})e^{it}),$$

where H has the same meaning as in Lemma 6. By this lemma,

$$(6.5) \qquad \int_{-\infty}^{+\infty} |H\tilde{g}|^p \, du \leq A_p^p \int_{-\infty}^{+\infty} |\tilde{g}|^p \, du.$$

Since, clearly, there is a $t_0(z)$ at which $|H\tilde{g}(\operatorname{re}(ze^{-it}))|$ attains its maximum, (6.4), (6.3), and (6.5) yield

$$\iint_D |S^*g|^p dxdy \le (\tfrac{1}{2}\pi A_p)^p \int_{-\infty}^{+\infty}\int_{-\infty}^{+\infty} |\tilde{g}(u)|^p du dv, \quad u+iv = z\exp[-it_0(z)],$$

whence, by $\tilde{g}(u) = g(z)$, (6.2) follows.

S t e p <u>C</u>. <u>We are going to prove the formulae</u>

$$(6.6)\quad S^*g(z) = (1/\pi)\lim_{\varepsilon\to 0+}\iint_{D(z;\varepsilon)} g_\zeta(\zeta)(1/|\zeta - z|)d\xi d\eta,$$

$$(6.7)\quad S^*g(z) = \frac{1}{\pi}\frac{\partial}{\partial z}\lim_{\varepsilon\to 0+}\iint_{D(z;\varepsilon)} g(\zeta)\left(\frac{1}{|\zeta - z|} - \frac{1}{|\zeta|}\right)d\xi d\eta,$$

for g as in Step B, including (6.3), though the proof given below remains valid for C^2 replaced with C^1.

To this end we observe that

$$S^*g(z) = (-1/\pi)\lim_{\varepsilon\to 0+}\iint_{D(z;\varepsilon)} g(\zeta)(1/|\zeta - z|)_\zeta\, d\xi d\eta,$$

so, by (6.3) and the first of Green's formulae (3.2) applied to $h^*(\zeta) = g(\zeta)/|\zeta - z|$ and $D^*(z;\zeta) = \operatorname{int}\Delta^{1/\varepsilon}(z)\setminus\Delta^\varepsilon(z)$, where $\Delta^{1/\varepsilon}(z)\supset D$, we obtain

$$S^*g(z) = \frac{1}{\pi}\lim_{\varepsilon\to 0+}\iint_{D(z;\varepsilon)} g_\zeta(\zeta)\frac{d\xi d\eta}{|\zeta - z|} + \frac{1}{2\pi i}\lim_{\varepsilon\to 0+}\iint_{D^*(z;\varepsilon)} g(\zeta)\frac{d\bar{\zeta}}{|\zeta - z|}.$$

Since $\operatorname{spt} g\subset D$, the second addend vanishes, and this proves (6.6). On the other hand,

$$\iint_{D(z;\varepsilon)} g_\zeta(\zeta)\frac{d\xi d\eta}{|\zeta - z|} = \iint_{D(0;\varepsilon)} g_\zeta(\zeta + z)\frac{d\xi d\eta}{|\zeta|} = \iint_{D(0;\varepsilon)} g_z(\zeta + z)\frac{d\xi d\eta}{|\zeta|}$$

$$= \frac{\partial}{\partial z}\iint_{D(0;\varepsilon)} g(\zeta + z)\frac{d\xi d\eta}{|\zeta|} = \frac{\partial}{\partial z}\iint_{D(z;\varepsilon)} g(\zeta)\frac{d\xi d\eta}{|\zeta - z|}.$$

Therefore (6.7) follows.

S t e p <u>D</u>. <u>We next proceed to derive the formula</u>

$$(6.8)\quad Sg = -S^*S^*g$$

for g as in Step B, including (6.3).

To this end we first apply (6.7) to the right-hand side of (6.8):

$$S^*S^*g(s) = \frac{1}{\pi}\frac{\partial}{\partial s}\lim_{\varepsilon\to 0+}\iint_{D(s;\varepsilon)} S^*g(\zeta)\left(\frac{1}{|\zeta-s|}-\frac{1}{|\zeta|}\right)d\xi d\eta .$$

Further, by (6.6), we have

$$S^*S^*g(s) = \frac{1}{\pi^2}\frac{\partial}{\partial s}\lim_{\varepsilon\to 0+}\iint_{D(s;\varepsilon)}\lim_{\delta\to 0+}\iint_{D(\zeta;\delta)} g_z h\,dxdy\,d\xi d\eta ,$$

where $h(z,\zeta,s) = |z-\zeta|^{-1}(|\zeta-s|^{-1}-|\zeta|^{-1})$. By the first of Green's formulae (3.2) applied to $h^*(z) = g(z)h(z,\zeta,s)$, we get

$$\lim_{\delta\to 0+}\iint_{D(\zeta;\delta)} g_z h\,dxdy = -\lim_{\delta\to 0+}\iint_{D(\zeta;\delta)} gh_z dxdy$$

$$-\frac{1}{2i}\lim_{\delta\to 0+}\int_{\partial D(\zeta;\delta)} gh\,d\bar{z}.$$

Since $\operatorname{spt} g \subset D$, the second addend vanishes. Thus, after interchanging the order of the improper integrals, we obtain

(6.9) $$S^*S^*g(s) = -\frac{1}{\pi^2}\frac{\partial}{\partial s}\lim_{\delta\to 0+}\iint_{D(\zeta;\delta)} g\lim_{\varepsilon\to 0+}\iint_{D(s;\varepsilon)} h_z d\xi d\eta\,dxdy.$$

In turn we observe that, by (6.3), (6.9) remains valid if we replace $D(s;\varepsilon)$ by $\mathbb{C}\setminus\Delta^{\varepsilon}(s)$, and that

$$\iint_{|\zeta-s|\geq\varepsilon} h_z d\xi d\eta = \frac{\partial}{\partial z}\iint_{|\zeta-s|\geq\varepsilon} h\,d\xi d\eta ,$$

(6.10) $$\lim_{\varepsilon\to 0+}\frac{\partial}{\partial z}\iint_{|\zeta-s|\geq\varepsilon} h\,d\xi d\eta = \frac{\partial}{\partial z}\lim_{\varepsilon\to 0+}\iint_{\varepsilon\leq|\zeta-s|\leq 1/\varepsilon} h\,d\xi d\eta .$$

Hence (6.9) becomes

(6.11) $$S^*S^*g(s) = (-1/\pi^2)(\partial/\partial s)\lim_{\delta\to 0+}\iint_{D(\zeta;\delta)} g\,\mathfrak{l}\,dxdy,$$

where

$$\mathfrak{l}(z,s) = \lim_{\varepsilon\to 0+}\frac{\partial}{\partial z}\iint_{\varepsilon\leq|\zeta-s|\leq 1/\varepsilon}\frac{1}{|z-\zeta|}\left(\frac{1}{|\zeta-s|}-\frac{1}{|\zeta|}\right)d\xi d\eta .$$

Since $\lim_{\varepsilon\to 0+}(\partial/\partial z)\{\iint_{\varepsilon\leq|\zeta-s|\leq 1/\varepsilon} - \iint_{\varepsilon\leq|\zeta|\leq 1/\varepsilon}\}(1/|\zeta||z-\zeta|)d\xi d\eta = 0$, we have

$$(6.12)\quad \mathfrak{I}(z,s) = \lim_{\varepsilon\to 0+} \frac{\partial}{\partial z} \iint\limits_{\varepsilon\le|\zeta-s|\le 1/\varepsilon} \frac{d\xi d\eta}{|\zeta - s||z-\zeta|} - \lim_{\varepsilon\to 0+} \frac{\partial}{\partial z} \iint\limits_{\varepsilon\le|\zeta|\le 1/\varepsilon} \frac{d\xi d\eta}{|\zeta||z-\zeta|} .$$

Changing the variables of integration according to the formulae $\zeta - s = (z-s)w$ in the first integral and $\zeta = zw$ in the second one, we get, by (1.5),

$$(6.13)\quad \mathfrak{I}(z,s) = \mathfrak{I}^*(z,s) - \mathfrak{I}^*(z,0),$$

where, on putting $u = \operatorname{re} w$, $v = \operatorname{im} w$,

$$\mathfrak{I}^*(z,s) = \lim_{\varepsilon\to 0+} \frac{\partial}{\partial z} \iint\limits_{\varepsilon\le|w|/|z-s|\le 1/\varepsilon} \frac{du dv}{|w||1-w|}$$

$$= \lim_{\varepsilon\to 0+} \frac{\partial}{\partial z} \int_0^{2\pi} \int_{\varepsilon/|z-s|}^{1/\varepsilon|z-s|} \frac{dr dt}{|1-re^{it}|}$$

$$= \lim_{\varepsilon\to 0+} (\partial/\partial z) \int_0^{2\pi} [(\varepsilon|z-s||1-e^{it}/\varepsilon|z-s||\sin 2t)^{-1}$$

$$- (\varepsilon^{-1}|z-s||1-\varepsilon e^{it}/|z-s||\sin 2t)^{-1}]dt$$

$$= -\tfrac{1}{2}(z-s)^{-1} \lim_{\varepsilon\to 0+} \int_0^{2\pi} [(\varepsilon|z-s||1-e^{it}/\varepsilon|z-s||)^{-1}$$

$$- (\varepsilon^{-1}|z-s||1-\varepsilon e^{it}/|z-s||)^{-1}]dt$$

$$= -\tfrac{1}{2}(z-s)^{-1} \int_0^{2\pi} dt = -\pi(z-s).$$

Consequently, by (6.11), (6.12), and (6.13), we obtain (cf. (4.5)):

$$S^*S^*g(s) = \frac{1}{\pi} \frac{\partial}{\partial s} \lim_{\delta\to 0+} \iint\limits_{D(\zeta;\delta)} g(z)\left(\frac{1}{z-s} - \frac{1}{z}\right)dxdy = -\frac{\partial}{\partial s} Tg(s).$$

Applying finally Lemma 5, we derive (6.8), as desired.

In this place it is natural to ask for the reason of keeping the term $g(\zeta)/|\zeta|$ in the integrand of $S^*g(z)$ in (6.7), though it could be deleted at all stages of our considerations without any influence on the final result (6.8). The answer is that in order to justify

(6.10) we have to check that

$$(\partial/\partial z)\iint_{|\zeta-s|\geq 1/\varepsilon} h\,d\xi\,d\eta \rightrightarrows 0 \quad\text{as}\quad \varepsilon\to 0+,$$

but this is not the case if we replace $h(z,\zeta,s)$ with $1/|z-\zeta||\zeta-s|$.

Step E. By (6.2) and Step A, S^* can be extended to a linear operator which associates with every L^p-function $g: D\to\mathbb{C}$, $1<p<+\infty$, the function S^*g of the same class and satisfies (6.2). Therefore, by (6.8), S can also be extended to a linear operator which associates with every L^p-function $g: D\to\mathbb{C}$, $1<p<+\infty$, the function Sg of the same class and satisfies $\|Sg\|_p\leq(\frac{1}{2}\pi A_p)^2\|g\|_p$. Hence $\|S\|_p\leq(\frac{1}{2}\pi A_p)^2<+\infty$, i.e. we conclude the first part of (6.1). Unfortunately our estimate is too rough to conclude the second part: Lemma 6 says that $A_2=1$ and hence we get $\|S\|_2\leq(\frac{1}{2}\pi)^2$ only. In order to improve this estimate we utilize Lemma 7, by which $\|S\|_p$ is continuous w.r.t. p and, consequently, by Step A, <u>we have only to demonstrate that</u>

$$\iint_D |Sg|^2dxdy=\iint_D |g|^2dxdy \tag{6.14}$$

for all C^2-functions $g: D\to\mathbb{C}$ with $\operatorname{spt} g\subset D$.

By Lemmas 5 and 4, we have

$$\iint_D |Sg|^2dxdy=\iint_D (Tg)_z\overline{(Tg)_z}dxdy$$

and

$$\iint_D |g|^2dxdy=\iint_D (Tg)_{\bar z}\overline{(Tg)_{\bar z}}dxdy,$$

respectively. Therefore, by (1.2) and both Green's formulae (3.2) applied to $h^*=(Tg)_{\bar z}\overline{Tg}|\Delta^{1/\varepsilon}$ and $h^*=(Tg)_z\overline{Tg}|\Delta^{1/\varepsilon}$, respectively, where $\Delta^{1/\varepsilon}\supset\operatorname{spt} g$, we get

$$\iint_D |g|^2dxdy=-\lim_{\varepsilon\to 0+}\iint_{|z|\leq 1/\varepsilon}(Tg)_{\bar z z}\overline{Tg}\,dxdy$$

$$+\tfrac{1}{2}i\lim_{\varepsilon\to 0+}\int_{|z|=1/\varepsilon}(Tg)_{\bar z}\overline{Tg}\,d\bar z$$

and

$$\iint_D |Sg|^2dxdy=-\lim_{\varepsilon\to 0+}\iint_{|z|\leq 1/\varepsilon}(Tg)_{z\bar z}\overline{Tg}\,dxdy$$

$$-\tfrac{1}{2}i\lim_{\varepsilon\to 0+}\int_{|z|=1/\varepsilon}(Tg)_z\overline{Tg}\,dz,$$

respectively, where, since, by Lemma 4, $(Tg)_{\bar z} = g$, then the integrals on the right-hand side of the first relation are independent of ε and the second of them vanishes for ε such that $\Delta^{1/\varepsilon} \supset \operatorname{spt} g$. Thus, since g is C^2, in order to conclude (6.14) it remains to prove that

$$\int_{|z|=1/\varepsilon} (Tg)_z \overline{Tg}\, dz \to 0 \quad \text{as} \quad \varepsilon \to 0+.$$

Owing to Lemma 5 we have $(Tg)_z = Sg$ and the desired relation follows from the estimate

$$(|z| - 1/\varepsilon^*)\max\{|Tg(z)|,\ (|z| - 1/\varepsilon^*)Sg(z)|\} \le (1/\pi)\iint_D |g|\, dxdy,$$

valid for $\Delta^{1/\varepsilon^*} \supset \operatorname{spt} g$, and this estimate is itself a direct consequence of the definitions of S and T. In this way the proof of (6.1) is completed.

Step F. _It remains to prove that_ (4.5) _implies_ (4.9) _for any_ L^p-_function_ $g: D \to ¢$, $p > 1$. Thus (cf. (3.1)) we have to verify that

$$(6.15) \qquad \iint_D (Sg)h\, dxdy = -\iint_D (Tg)h_z dxdy$$

for all C^1-functions $h: D \to ¢$ with compact support. To this end we take a sequence of C^2-functions $g_n: D \to ¢$ with $\operatorname{spt} g_n \subset D$, such that $\|g_n - g\|_p \to 0$, what is justified by Step A. By Lemma 5, (6.15) holds with g_n substituted for g, so we have only to show that

$$(6.16) \qquad \iint_D (Sg_n - Sg)h\, dxdy \to 0, \qquad \iint_D (Tg_n - Tg)h_z dxdy \to 0.$$

The first relation follows, by Hölder's inequality, from $\|Sg_n - Sg\|_p \to 0$, which is itself a direct consequence of (6.1). In order to obtain the second one we observe that, by Hölder's inequality,

$$|T(g_n - g)(z)| \le (1/\pi)\Big(\iint_D |\zeta - z|^{-p'} d\xi d\eta\Big)^{1/p'} \|g_n - g\|_p$$

$$= (1/\pi)\Big(\iint_{D(z)} |\zeta - 1|^{-p'} d\xi d\eta\Big)^{1/p'} |z|^{1-2/p} \|g_n - g\|_p,$$

where $1/p + 1/p' = 1$, $D(z) = \{\zeta:\ z\zeta \in D\}$, and $z \ne 0$ (for $z = 0$ we may get an analogous estimate with $|\zeta - 1|^{-p'}$ and $|z|^{1-2/p}$ replaced by $|\zeta|^{-p'}$ and 1, respectively). Since h has compact support, this suffices to conclude the proof of (6.16) and, consequently, of (6.15). In this way the proof of Lemma 8 is completed.

COROLLARY 1. $\|S\|_p$ *is continuous (w.r.t. p), strictly decreasing for* $1<p\leq 2$, *strictly increasing for* $2\leq p<+\infty$, *grows to* $+\infty$ *as* $p \longrightarrow 1+$ *or* $p \longrightarrow +\infty$, *and satisfies conditions*

$$(6.17)\quad \limsup_{p\to 1+}\{\|S\|_p/(p-1)^2\} = c_1 < +\infty, \quad \limsup_{p\to +\infty}(\|S\|_p/p^2) = c_2 < +\infty .$$

Proof. As we have already remarked in Step E of the preceding proof, continuity of $\|S\|_p$ is a direct consequence of Lemma 7. By the same lemma, in particular, inequality (5.1), $\|S\|_{p_1} \leq \|S\|_2^{1-t}\|S\|_{p_2}^{t}$, where $t=(\frac{1}{2}-1/p_1)/(\frac{1}{2}-1/p_2)$, $p_2>p_1>2$. Since $t<1$ and, by Lemma 8, $\|S\|_2=1$, then $\|S\|_p$ is strictly increasing for $2\leq p<+\infty$ and grows to $+\infty$ as $p\longrightarrow+\infty$. In a similar way we manage with the other case. Finally, conditions (6.17) follow directly from the inequality $\|S\|_p \leq (\frac{1}{2}\pi A_p)^2$, derived in Step E of the preceding proof, and from the formula $A_p = A_{p'} = [(1-1/p)^{-2/p}-1]^{-\frac{1}{2}}$, where $p\geq 2$ and $1/p+1/p'=1$, obtained in Section 5.

It is interesting to remark that Calderón and Zygmund [2] proved stronger estimates

$$\limsup_{p\to 1+}\{\|S\|_p/(p-1)\} = \tilde{c}_1 < +\infty, \quad \limsup_{p\to +\infty}(\|S\|_p/p) = \tilde{c}_2 < +\infty .$$

7. A Tricomi singular integral equation and integrability of derivatives

Thereafter we often use the phrase "$w=f(z)$ is a solution of $\Phi(w,\ldots)=0$" which is the same as "$w=f(z)$ satisfies $\Phi(w,\ldots)=0$".

According to Lemma 4, $w=f(z)$ is a solution of (4.3) iff it is of the form (4.5). Now, by means of Lemma 8, we can reduce the discussion of the Beltrami differential equation (4.2), which has been replaced by the system of (4.3) and (4.4), to the discussion of a *Tricomi singular integral equation* (cf. Tricomi [1]), namely

$$(7.1)\quad \omega - \mu S\omega = \mu\Phi' \quad \text{a.e.}$$

This is given by the following

LEMMA 9. Suppose that μ *is a measurable function with compact support, defined in a domain* $D\subset \mathbb{C}$, *and such that* $\|\mu\|_\infty<1$. *Then* $w=f(z)$, *given by* (4.5), *with* g *locally* L^p, $p>1$, *and* Φ *holomorphic, is a solution of* (4.2) *or – equivalently* – (4.3) *and* (4.4), *iff* $\omega = g(z)$ *is a solution of* (7.1).

Proof. If $w=f(z)$ is a solution of (4.2) or – equivalently –

(4.3) and (4.4), then this fact, together with relation (4.9) which, by Lemma 8, is a consequence of (4.5), yield (7.1) with $\omega = g(z)$. Conversely, if $\omega = g(z)$ is a solution of (7.1), then, since $\mathrm{spt}\,\mu$ is compact, $\mathrm{spt}\,g$ is compact as well. On the other hand, g is locally of a class L^p, $p > 1$, so, by Lemma 4, $w = f(z)$ satisfies (4.3). But, by Lemma 8, (4.5) implies (4.9). Consequently, relations (4.3) with $w = f(z)$, (4.9), and (7.1) with $\omega = g(z)$ yield (4.4) and (4.2) with $w = f(z)$, thus concluding the proof.

We notice that the compactness of $\mathrm{spt}\,\mu$ implies that, in the above lemma, g is L^p, so – finally – we have reduced the problem in question to studying L^p-solutions of (7.1). Furthermore, we confine ourselves to the case where $D = \mathbb{C}$ and $\Phi = \mathrm{id}_{\mathbb{C}}$, what will be justified in Section 8. We have

LEMMA 10. Suppose that μ is a measurable function with compact support, defined in $\mathbb{C}$ and such that $\|\mu\|_\infty < 1$. Then there exists a solution $\omega = g(z)$ of (7.1) with $\Phi = \mathrm{id}_{\mathbb{C}}$ which is L^p for $p \in [1; 2+\infty)$, $\varepsilon > 0$ (more exactly, it is sufficient to assume that $\|\mu\|_\infty \|S\|_{2+\varepsilon} \le 1$), and if $\omega = \hat{g}(z)$ is another such solution, then $\hat{g} = g$ a.e. Moreover, g has compact support.

Proof. Consider the sequence (S_n) of operators defined by the formulae $S_n\mu = S(\mu S_{n-1}\mu)$, where $S_o\mu = 1$ identically. Hence

$$(7.2)\qquad g_n = \mu + \mu S g_{n-1},$$

where $g_{n-1} = \mu S_o\mu + \ldots + \mu S_{n-1}\mu$. By Lemma 8,

$$\|S_n\mu\|_p \le \|S\|_p \|\mu S_{n-1}\mu\|_p \le \|S\|_p \|\mu\|_\infty \|S_{n-1}\mu\|_p, \quad \|S\|_p < +\infty.$$

On the other hand, $\|S_o\mu\|_p = |\mathrm{spt}\,\mu|^{1/p}$, whence $\|S_n\mu\|_p \le (\|S\|_p\|\mu\|_\infty)^n \times |\mathrm{spt}\,\mu|^{1/p}$. Therefore, by the definition of (g_n), we obtain

$$\|g_m - g_n\|_p \le \|\mu\|_\infty \sum_{j=m+1}^{n} \|S_j\mu\|_p \le \|\mu\|_\infty |\mathrm{spt}\,\mu|^{1/p} \frac{(\|\mu\|_\infty \|S\|_p)^{m+1}}{1 - \|\mu\|_\infty \|S\|_p}$$

$$\longrightarrow 0 \text{ as } m, n \longrightarrow +\infty,\ m < n,\ \|\mu\|_\infty \|S\|_p < 1.$$

By Lemma 8, $\|S\|_2 = 1$. On the other hand, by Lemma 7, $\|S\|_p$ is continuous, and hence there is an $\varepsilon > 0$ such that $\|\mu\|_\infty \|S\|_p < 1$ for $|p-2| < \varepsilon$. Since L^p is complete, there exists an L^p-function g such that $\|g - g_n\|_p \longrightarrow 0$ as soon as $|p-2| < \varepsilon$.

Next, by Lemma 8 again, we conclude that $\|Sg_n - Sg\|_p \longrightarrow 0$ for

$|p-2|<\varepsilon$. Now we utilize a well known theorem that enables to conclude from the convergence w.r.t. $\| \ \|_p$ of a sequence the pointwise convergence a.e. of a subsequence of it (cf. e.g. Zygmund [1], p. 73). Therefore there is a subsequence (g_{n_k}) such that $g_{n_k} \to g$ a.e. and $Sg_{n_k} \to Sg$ a.e. Consequently, from (7.2) we infer that $\omega = g(z)$ is a solution of (7.1) with $\Phi = \mathrm{id}_{\mathbb{C}}$ and we already know that it is L^p for $|p-2|<\varepsilon$. Thus, since $\operatorname{spt}\mu$ is compact, $\operatorname{spt} g$ is compact as well, whence – on applying Hölder's inequality – we infer that g is L^p for $p \in [1; 2+\varepsilon)$. Finally, if $\omega = \hat{g}(z)$ were another such solution of (7.1) with $\Phi = \mathrm{id}_{\mathbb{C}}$, and $g - \hat{g} = 0$ a.e., then $\|g-\hat{g}\|_2 \le \|\mu\|_\infty \|Sg - S\hat{g}\|_2$ would imply, by Lemma 8, that $\|\mu\|_\infty \ge \|S\|_2^{-1} = 1$, what contradicts $\|\mu\|_\infty < 1$, thus concluding the proof.

Bojarski [1, 2] observed that Lemma 3 admits the following interesting extension:

<u>THEOREM 1</u>. <u>Suppose that</u> $f\colon D \to D' \subset \mathbb{C}$, <u>where</u> D <u>is a domain in</u> $\mathbb{C}$, <u>is a</u> Q-<u>qc homeomorphism</u>. <u>Then</u> f <u>possesses distributional</u> L^p-<u>derivatives</u> f'_z <u>and</u> $f'_{\bar{z}}$ <u>for all</u> $p \in [1; 2+\varepsilon)$, <u>where</u> ε <u>is positive, depends on</u> Q <u>only, and tends to</u> $+\infty$ <u>as</u> $Q \to 1+$.

In this section we confine ourselves to proving Theorem 1 for $D = D' = \mathbb{C}$ and $f'_{\bar{z}}/f'_z$ with compact support. The first restriction can be avoided by means of introductory considerations in Section 8, before Theorem 2. The second restriction can be avoided by means of the same considerations and by Theorem 2 applied to the complex dilatation equal $f'_{\bar{z}}(z)/f'_z(z)$ for $z \in D \setminus \{z\colon |z| \ge 1/\hat{\varepsilon}\}$ and equal 0 otherwise, and letting $\hat{\varepsilon} \to 0+$. This is correct since we do not apply Theorem 1 in Section 8. One can naturally ask why we do not formulate Theorem 1 after Theorem 2. The reason is that the main idea of Theorem 1 is very close to that of Lemmas 9 and 10, and we do not apply Theorem 1 later on, whereas we utilize Theorem 2 frequently. Thus, since – by means of Lemmas 1 and 3 – f possesses distributional L^p-derivatives f'_z and $f'_{\bar{z}}$ (where we may take $p=2$), we finally confine ourselves to proving

<u>LEMMA 11</u>. <u>Suppose that</u> $f\colon \mathbb{C} \to \mathbb{C}$ <u>is a</u> Q-<u>qc homeomorphism, whose distributional</u> L^p-<u>derivatives are such that</u> $f'_{\bar{z}}/f'_z$ <u>has compact support</u>. <u>Then as</u> p <u>we may take any number within</u> $[1; 2+\varepsilon)$, <u>where</u> ε <u>is positive, depends on</u> Q <u>only, and tends to</u> $+\infty$ <u>as</u> $Q \to 1+$.

<u>Proof</u>. By Lemma 3, $w = f(z)$ is a solution of (4.2), where μ is measurable in $\mathbb{C}$ and

(7.3) $\|\mu\|_\infty \leq (Q-1)/(Q+1)$.

Hence, by Lemma 4, f can be expressed in the form (4.5), where $g = f'_{\bar z}$ and Φ is a holomorphic function continuable to a function meromorphic in $\mathbb{E}$ and having the only (simple) pole at ∞. Hence we may, without any loss of generality, assume that $\Phi = \mathrm{id}_{\mathbb{C}}$. Applying now Lemma 9 we see that $\omega = f'_{\bar z}(z)$ is a solution of (7.1). Next, by Lemma 10, we conclude that $f'_{\bar z}$ is L^p for all $p \in [1; 2+\varepsilon)$, where ε fulfils the condition $\|\mu\|_\infty \|S\|_{2+\varepsilon} \leq 1$, whence, by (7.3), it is sufficient to assume that

(7.4) $\|S\|_{2+\varepsilon} \leq (Q+1)/(Q-1)$.

Finally, utilizing Lemma 8, we obtain, by $\Phi = \mathrm{id}_{\mathbb{C}}$, that $\|f'_z - 1\|_p = \|Sf'_{\bar z}\|_p \leq \|S\|_p \|f'_{\bar z}\|_p$, thus, by Minkowski's inequality, concluding the proof.

The supremum of all admissible numbers ε for the class of all Q-qc mappings, Q fixed, equals $2Q/(Q-1)$, as proved by Gol'dštein [1]. The proof is based on an estimation of the value distribution for the Jacobian of the mapping. The example $f(z) = z|z|^{1/Q-1}$ for z in a neighbourhood of 0 shows that $\sup \varepsilon \leq 2Q/(Q-1)$. Already Lehto [1] proved that $\sup \varepsilon \geq 2Q^{2/c}/(Q^{2/c}-1)$, where $c = \lim\inf[(Q-1)\sup\varepsilon]$ as $Q \to 1+$ and posed a conjecture that $c = 2$, whose positive solution would imply the equality of Gol'dštein.

Another interesting question is to find an analogue of Theorem 1 for higher dimensions and – unfortunately – the elegant Bojarski's proof does not suggest, what the situation is in that case. This problem has been solved not long ago by Gehring [5], whose proof works also in the two-dimensional case.

8. A special case of the theorem on existence and uniqueness

We proceed to prove now a special case of the theorem on existence and uniqueness of qc homeomorphisms with a preassigned complex dilatation. We essentially follow Ahlfors [5], but the original result goes back, under different hypotheses to Gauss [1], Lichtenstein [1], Lavrentieff [1], and Vekua [1, 2].

We begin with the following (Ahlfors [5], pp. 95-96):

LEMMA 12. *If an integrable function Φ with compact support (in $\mathbb{C}$) possesses distributional L^p-derivatives Φ'_z and $\Phi'_{\bar z}$, then $(\exp\Phi)'_z = (\exp\Phi)\Phi'_z$ and $(\exp\Phi)'_{\bar z} = (\exp\Phi)\Phi'_{\bar z}$ a.e.*

Proof. Let $\operatorname{spt}\Phi\subset\Delta^r$, say. We apply the fact that the class of all C^1-functions f^* with $\operatorname{spt} f^*\subset\Delta^r$ is dense in the class of all functions f locally L^p with $\operatorname{spt} f\subset\Delta^r$; the distance of f and f^* being defined by $\|f-f^*\|_p$, p arbitrary in $[1;+\infty)$ (cf. e.g. Lehto and Virtanen [1], p. 149, or [2], p. 142). Hence there is a sequence (Φ_n) of C^1-functions such that $\operatorname{spt}\Phi_n\subset\Delta^r$, $\Phi_n\to\Phi$ a.e., and $\|\Phi'_z - \Phi'_{nz}\|_p\to 0$. Consequently, if we extend all the functions in question to $\operatorname{cl}\Delta^r$ by the value 0, then by the first of Green's formulae (3.2) applied to $h^*=(\exp\Phi_n)h$ and $D^*=\operatorname{int}\Delta^r$, where h is an arbitrary C^1-function with compact support (in D^*), we get

$$\iint_{\Delta^r}(\exp\Phi_n)\Phi'_{nz}h\,dxdy = -\iint_{\Delta^r}(\exp\Phi)h_z dxdy,$$

whence, by letting $n\to+\infty$, we arrive at the desired result for Φ'_z, the proof for $\Phi'_{\bar z}$ being analogous.

We then proceed to prove the announced

LEMMA 13. *Suppose that μ is a continuous function with compact support (in $\mathbb{C}$), defined in $\mathbb{E}$, bounded by a constant $q<1$, and such that μ'_z exists and is bounded by a constant q. Consider the equations (4.5) and (7.1) with $\Phi=\mathrm{id}$. Then the function $w=f(z)$, given by (4.5), where $\omega=g(z)$ is the unique solution of (7.1), is a solution of (4.2) which represents a regular Q-qc homeomorphism with a fixpoint at ∞, where $Q=(1+q)/(1-q)$.*

Proof. The fact that $w=f(z)$ is a solution of (4.2), or – equivalently – (4.3) and (4.4), is a direct consequence of Lemma 9. We now proceed to prove that one can choose $\log\hat g$ in $\mathbb{E}$, where $\mu\hat g = g$, so that it is continuous and possesses distributional L^p-derivatives $(\log\hat g)'_z$ and $(\log\hat g)'_{\bar z}$ which satisfy

$$(\log\hat g)'_{\bar z} = \mu(\log\hat g)'_z + \mu'_z \quad \text{a.e.}$$

To this end we consider the equation $\omega = S(\mu\omega+\mu'_z)$. In order to solve it we further consider the sequence (S_n) of operators defined by the formulae $S_n\mu = S(\mu S_{n-1}\mu)$, where $S_o\mu = S\mu'_z$. Hence $g_n = S(\mu g_{n-1} + \mu'_z)$, where $g_{n-1} = S_o\mu+\ldots+S_{n-1}\mu$. By Lemma 8,

$$\|S_n\mu\|_p\le\|S\|_p\|\mu S_{n-1}\mu\|_p\le\|S\|_p\, q\|S_{n-1}\mu\|_p,\quad \|S\|_p<+\infty.$$

On the other hand $\|S_o\mu\|_p\le\|S\|_p\, q'|\operatorname{spt}\mu|^{1/p}$, whence $\|S_n\mu\|_p\le\|S\|_p^{n+1}\times q^n q'|\operatorname{spt}\mu|^{1/p}$. Therefore, by the definition of (g_n), we obtain

$$\|g_m - g_n\|_p \le \sum_{j=m+1}^{n} \|S_j\mu\|_p \le q'|\operatorname{spt}\mu|^{1/p}\|S\|_p(q\|S\|_p)^{m+1}/(1-q\|S\|_p)$$

$$\longrightarrow 0 \text{ as } m,n \longrightarrow +\infty,\quad m<n,\quad q\|S\|_p<1.$$

Arguing as in the proof of Lemma 10 we conclude that the equation in question has a unique (in an analogous sense) solution g_∞ which is L^p for $p \in [1;2+\varepsilon)$, $\varepsilon > 0$, where ε has to satisfy $q\|S\|_{2+\varepsilon} \le 1$. Consequently, the function $\tilde{g}$, defined in $\mathbb{E}$ by the formula

$$\tilde{g}(z) = [T(\mu g_\infty + \mu_z')](z) - [T(\mu g_\infty + \mu_z')](\infty),$$

is continuous and, by Lemmas 4 and 8, possesses distributional L^p-derivatives

$$\tilde{g}_{\bar{z}}' = \mu g_\infty + \mu_z', \quad \tilde{g}_z' = S(\mu g_\infty + \mu_z') = g_\infty.$$

Therefore, by Lemma 12 and the uniqueness of g_∞, we obtain $\hat{g} = \exp\tilde{g}$, what suffices to conclude the desired statement.

The statement proved above and Lemma 12 show that $\hat{g}$ is continuous and possesses distributional L^p-derivatives which satisfy $\hat{g}_{\bar{z}}' = (\mu\hat{g})_z'$ a.e. Thus, since f is a solution of (4.3) and (4.4), f is C^1 and, since it is given by (4.5) with $\Phi = \mathrm{id}$, it remains to prove that f is a sense preserving homeomorphism.

To this end we observe that, firstly, by (1.5), $J_f = (1-|\mu|^2)\times \exp(2\operatorname{re}\tilde{g})$. Hence $\operatorname{sgn} J_f(z) = 1$ everywhere, so, as remarked at the beginning of Section 1, f is a local sense preserving homeomorphism. Secondly we observe that, by (4.5) with $\Phi = \mathrm{id}$ and the compactness of $\operatorname{spt}\mu$, f is holomorphic in a vicinity U of ∞ and $f(z) = z + o(1)$ as $z \longrightarrow \infty$. Hence f is one-to-one in some neighbourhood $U_0 \subset U \cup \{\infty\}$ of ∞. This yields

$$(1/2\pi i)\int_{\partial f[\Delta^r]} (w - w_0)^{-1}\,dw = 1$$

for all r such that $\operatorname{fr}\Delta^r \subset U_0$ and $w_0 \in f[\Delta^r]$. Now, the following theorem holds (cf. e.g. Goursat [1], vol. I, pp. 388-389). Let D be a Jordan domain with smooth boundary ∂D and let $h\colon \operatorname{cl} D \longrightarrow \mathbb{C}$ be a C^1-function. Suppose that $h(z) \ne 0$ on $\operatorname{fr} D$ and $J_h(z) \ne 0$ at zeros of h (which are, consequently, isolated). If P and N are numbers of zeros of h for which $J_h(z) > 0$ and $J_h(z) < 0$, respectively, then

$$(1/2\pi)\int_{\partial D} d(\arg h) = P - N.$$

In our case

$$(1/2\pi)\int_{\partial\Delta^r} d\{\arg[f(z)-w_0]\} = (1/2\pi i)\int_{\partial f[\Delta^r]} (w-w_0)^{-1}dw = 1$$

and $J_{f-w_0} > 0$. Hence $f - w_0$ has exactly one zero in Δ^r. This means that f is univalent in Δ^{r^o}, thus concluding the proof since we may take r arbitrarily large.

Finally we shall prove the chain rule for the distributional derivatives of composed homeomorphic solutions of Beltrami equations: $w=f(z)$, where $f=f_2\circ f_1$ and the corresponding complex dilatations are μ, μ_1, and μ_2, respectively. We are even able to weaken slightly the assumptions (for the sake of convenience we also change the notation):

LEMMA 14. Let $w=f(z)$ be a homeomorphic solution of (4.2) in an open set $D\subset\mathbb{E}$, where μ is a measurable function such that $\|\mu\|_\infty<1$. Further, let g possess distributional L^2-derivatives g'_z and $g'_{\bar z}$. Then $g\circ f$ has in its set of definition the generalized derivatives

(8.1) $(g\circ f)'_z = (g'_z\circ f)f'_z + (g'_{\bar z}\circ f)\bar f'_z$ a.e.,

(8.2) $(g\circ f)'_{\bar z} = (g'_z\circ f)f'_{\bar z} + (g'_{\bar z}\circ f)\bar f'_{\bar z}$ a.e.

Proof (cf. Ahlfors and Bers [1]). Formulae (8.1) and (8.2) certainly hold if f and g are C^1-functions. We have assumed that g'_z and $g'_{\bar z}$ are locally L^2. For f'_z and $f'_{\bar z}$ the same is valid by Lemmas 1 and 2. Therefore, by (3.1) and (3.2) it is possible to find sequences (f_m) and (g_n) of C^1-functions so that $f_m\to f$, $g_n\to g$, and

(8.3) $$\iint_D (|f_{mz}-f'_z|^2 + |f_{m\bar z}-f'_{\bar z}|^2)\,dxdy \to 0,$$

(8.4) $$\iint_D (|g_{nz}-g'_z|^2 + |g_{n\bar z}-g'_{\bar z}|^2)\,dxdy \to 0.$$

In fact, to this end it is sufficient to take

$$f_m(z) = f*\vartheta_m(z) = \iint_D f(w)\,\vartheta_m(z-w)\,dudv,\quad z\in D,$$

$$g_n(z) = g*\tau_n(z),\quad z\in f[D],$$

where $u=\operatorname{re} w$, $v=\operatorname{im} w$, and $\vartheta_m(z)=a_m\exp(|z-s|^2-n^{-1})$ for $|z-s|<1/n$, $\vartheta_m(z)=0$ for $|z-s|\ge 1/n$ with the integral of each ϑ_m over D being equal to 1 and s being an arbitrary (fixed) point of D, while τ_n are the analogues of ϑ_m with s replaced by an arbitrary (fixed) point of $f[D]$.

Consider now the integrals

$$I_1 = \iint_D |g_{nz} \circ f - g'_z \circ f| \, |f'_z| \, dxdy,$$

$$I_2 = \iint_D |g_{nz} \circ f - g_{nz} \circ f_m| \, |f'_z| \, dxdy,$$

$$I_3 = \iint_D |g_{nz} \circ f_m| \, |f'_z - f_{mz}| \, dxdy.$$

By the Schwarz inequality and (8.4)

$$I_1 \le \left(\iint_D |g_{nz} - g'_z|^2 \circ f \, dxdy \right)^{\frac{1}{2}} \left(\iint_D |f'_z|^2 dxdy \right)^{\frac{1}{2}} \to 0.$$

Similarly, by (8.3), $I_3 \to 0$ for any fixed n and the convergence $I_2 \to 0$ for any fixed n is obvious. Consequently, it is possible to choose m and n so that

$$\iint_D |(g_{nz} \circ f_m) f_{mz} - (g'_z \circ f) f'_z| \, dxdy$$

be arbitrarily small. The same holds for

$$\iint_D |(g_{n\bar{z}} \circ f_m) \bar{f}_{mz} - (g'_{\bar{z}} \circ f) \bar{f}'_z| \, dxdy,$$

and hence (8.1) follows. The proof of (8.2) is quite analogous.

9. Bojarski's proof for the general case

The theorem on existence and uniqueness of qc mappings with a preassigned measurable complex dilatation in the case of simply connected domains is due to Morrey [1]. Other proofs were given by Bers and Nirenberg [1], Bojarski [1, 2], Belinskiĭ and Pesin [1], and Dittmar [1]. We follow the method of Bojarski, at first, however, we make some preparatory considerations that explain a particular form of this theorem chosen by him.

Suppose that D and D' are conformally equivalent simply connected domains or closures of Jordan domains in $\mathbb{E}$, and μ a measurable function defined in D, such that $\|\mu\|_\infty < 1$. We are interested in finding homeomorphic solutions $w = f(z)$ of (4.2) which map D onto D'. In view of the classical theorems of Riemann and Osgood-Carathéodory, as well as of the formula

(9.1) $\mu = [\mu_1 + (\mu_2 \circ f_1)\exp(-2i \arg f'_{1z})]/[1 + \mu_1 (\mu_2 \circ f_1)\exp(-2i \arg f'_{1\bar{z}})]$

which relates a.e. the complex dilatation of $f_2 \circ f_1$ with those of f_1 and f_2, we may confine ourselves to the cases: (i) $D = D' = \operatorname{int} \Delta$, (ii) $D = D' = \Delta$. (iii) $D = D' = \mathbb{C}$, (iv) $D = D' = \mathbb{E}$. Formula (9.1) is an easy consequence of Lemmas 1, 2, and 14, especially of formulae (8.1) and (8.2).

Case (iii) can be transformed to (iv) on putting $f(\infty) = \infty$ since, as it can easily be verified, the extended mapping is qc. Case (i) can be transformed to (ii) since any Q-qc homeomorphism from $\operatorname{int} \Delta$ onto $\operatorname{int} \Delta$ can be continued to a Q-qc homeomorphism from Δ onto Δ, as it will follow from Corollary 2 proved at the end of this section.

Thus we are led to two important classes of Q-qc mappings: S_Q and S_Q^*. A function f is said to be of the class S_Q [resp. S_Q^*] if it maps Δ [resp. $\mathbb{E}$] onto itself Q-quasiconformally with $f(0) = 0$, $f(1) = 1$ [and $f(\infty) = \infty$]. The normalization conditions for S_Q as well as for S_Q^* give no loss of generality owing to the properties of homographies (i.e. fractional linear transformations).

We are going to formulate the theorem on existence and uniqueness at first for $f \in S_Q^*$.

THEOREM 2. Suppose that μ is a measurable function defined in $\mathbb{E}$ and such that $\|\mu\|_\infty < 1$. Then there exists a unique solution $w = f(z)$ of (4.2) which represents a homeomorphism belonging to S_Q^*, where

(9.2) $Q = (1 + \|\mu\|_\infty)/(1 - \|\mu\|_\infty)$.

P r o o f. For greater clarity the proof is divided into ten steps.

S t e p A. We claim that we may confine ourselves to the problem in question for μ with compact support (in $\mathbb{C}$). Indeed, let us write $\mu = \mu_1 + \mu_2$, where μ_2 has compact support and $\mu_1 | \operatorname{spt} \mu_2 = 0$. Consider the equation (4.2) with μ replaced by μ_1^* defined by $\mu_1^*(z) = e^{4i \arg z} \times \mu_1(1/\bar{z})$, $z \in \mathbb{E}$. If the problem were solved for the complex dilatations with compact supports, there would exist a unique solution $w = f_1^*(z)$ of this equation which would represent a homeomorphism of a class $S_{Q_1}^*$. Hence $w = \check{f}_1(z)$, given by $f_1(w) = 1/\overline{f_1^*(1/\bar{w})}$, $w \in \mathbb{E}$, would be the unique solution of (4.2) with μ replaced by some $\tilde{\mu}_1$, which represented a homeomorphism of $S_{Q_1}^*$. In order to calculate $\tilde{\mu}_1$ we observe that the complex dilatation of f_1 equals μ_1 a.e. and that the complex dilatation of id is 0. Hence, by (9.1), $\tilde{\mu}_1 = -(\mu_1 \circ \check{f}_1) \times \exp(-2i \arg \check{f}'_{1z})$. Consider next the equation (4.2) with μ replaced by

$$\mu_2^* = [\tilde{\mu}_1 + (\mu \circ \check{f}_1)\exp(-2i \arg \check{f}'_{1z})]/[1 + \bar{\tilde{\mu}}_1(\mu \circ \check{f}_1)\exp(-2i \arg \check{f}'_{1\bar{z}})].$$

If the problem were solved for complex dilatations with compact supports, there would exist a unique solution $w = f_2^*(z)$ of this equation which would represent a homeomorphism of a class $S_{Q_2}^*$. Hence, by (9.1), $w = f(z)$, given by $f = f_2^* \circ f_1$, would be the unique solution of (4.2) which represented a homeomorphism of S_Q^* with Q given by (9.2).

S t e p B. We now apply the fact that for any $r > 0$ and q, $0 \le q < 1$, the class of all C^1-functions μ^* with $\operatorname{spt} \mu^* \subset \Delta^r$ and $\|\mu\|_\infty \le q$ is dense in the class of all measurable μ with $\operatorname{spt} \mu \subset \Delta^r$ and $\|\mu\|_\infty \le q$; the distance of μ and μ^* being defined by $\|\mu - \mu^*\|_p$, p arbitrary in $[1; +\infty)$ (cf. e.g. Lehto and Virtanen [1], p. 149, or [2], p. 142). Hence, given μ, there is a sequence (μ_n) of C^1-functions which approximates μ in the above sense.

S t e p C. By Lemma 10, for each positive integer n there is a number $\varepsilon > 0$ such that the equation (7.1) with μ replaced by μ_n and $\Phi = \mathrm{id}$ has an L^p-solution $\omega = g_n(z)$ with $\operatorname{spt} g_n \subset \Delta^r$, where $|p - 2| < \varepsilon$, and since $\|\mu_n\|_\infty \le q$ for all n, we may take ε independent of n. Therefore, by Lemma 9, $w = f_n(z)$, given by $f_n = \mathrm{id} + Tg_n$, is a solution of (4.2) with μ replaced by μ_n. We claim that there is an L^p-function g with $\operatorname{spt} g \subset \Delta^r$ such that $\|g - g_n\|_p \longrightarrow 0$.

In fact, by (7.1) with μ replaced by μ_n, $\omega = g_n(z)$, and $\Phi = \mathrm{id}$, and by Lemma 8, we have

$$\|g_n\|_p < \|\mu_n\|_\infty \|S\|_p \|g_n\|_p + \|\mu_n\|_p \le q\|S\|_p\|g_n\|_p + q(\pi r^2)^{1/p}, \qquad \|S\|_p < +\infty,$$

since $\|\mu_n\|_\infty \le q$ and $\operatorname{spt} \mu_n \subset \Delta^r$. Hence, for p satisfying the conditions $q\|S\|_p < 1$ and $|p - 2| < \varepsilon$ (by Lemma 7, inequality $q\|S\|_p < 1$ implies $|p - 2| < \varepsilon$), we get

$$(9.3) \qquad \|g_n\|_p \le q(\pi r^2)^{1/p}/(1 - q\|S\|_p).$$

Further, from (7.1) with μ replaced by μ_m, $\omega = g_m(z)$, and $\Phi = \mathrm{id}$, we obtain

$$g_m - g_n = \mu_m S(g_m - g_n) + (\mu_m - \mu_n) S g_n + \mu_m - \mu_n,$$

whence, again by Lemma 8 and $\|\mu_m\|_\infty \le q$,

$$(1 - q\|S\|_p)\|g_m - g_n\|_p \le \|(\mu_m - \mu_n) S g_n\|_p + \|\mu_m - \mu_n\|_p.$$

On the other hand, by Hölder's inequality,

$$\|(\mu_m - \mu_n) S g_n\|_p \le \|\mu_m - \mu_n\|_{p\tilde{p}'} \cdot \|S g_n\|_{p\tilde{p}},$$

where $\tilde{p}>1$ and $1/\tilde{p}+1/\tilde{p}'=1$. Therefore, by (9.3), Lemma 8, and $\|\mu_m-\mu_n\|_p \to 0$, $\|\mu_m-\mu_n\|_{p\tilde{p}'} \to 0$ as $m,n \to +\infty$, we conclude that

$$(9.4)\quad \|g_m-g_n\|_p \le \frac{1}{1-q\|S\|_p}\{q(\pi r^2)^{1/p}\|\mu_m-\mu_n\|_{p\tilde{p}'}\frac{\|S\|_{p\tilde{p}}}{1-q\|S\|_{p\tilde{p}}} + \|\mu_m-\mu_n\|_p\} \to 0 \text{ as } m,n\to\infty,$$

provided that $q\|S\|_p<1$, $|p-2|<\varepsilon$, and we have chosen $\tilde{p}$ so close to 1 that $q\|S\|_{p\tilde{p}}<1$. Finally we apply a well known theorem on weak convergence of L^p-functions which says that there is an L^p-function g such that $\|g-g_n\|_p \to 0$ iff $\|g_m-g_n\|_p \to 0$ as $m,n\to+\infty$, whence in our case $\|g-g_n\|_p \to 0$ indeed.

S t e p D. Let f be defined by (4.5), where $\Phi = \mathrm{id}$ and $D=\mathbb{E}$. For the sake of convenience in notation we have denoted the above function by f, although we are going to prove that not actually f, but some related function $af+b$ belongs to S_Q^*, where a and b are certain constants. We now proceed to prove that $f_n \rightrightarrows f$. To this end we apply relations $f_m = \mathrm{id}+Tg_m$, $f_n=\mathrm{id}+Tg_n$ and $\operatorname{spt}\mu_m \cup \operatorname{spt}\mu_n \subset \Delta^r$. Thus, for $s\in\Delta^{r*}\supset\Delta^r$, by the definition of T and Hölder's inequality, we have

$$\begin{aligned}|f_m(s)-f_n(s)| &\le (1/\pi)\iint_{\Delta^{r*}} |z-s|^{-1}|g_m-g_n(z)|\,dxdy\\ &\le (1/\pi)\|g_m-g_n\|_p\Big(\iint_{\Delta^{r*}} |z-s|^{-p'}dxdy\Big)^{1/p'}\\ &\le (1/\pi)\|g_m-g_n\|_p\Big(\iint_{\Delta^{2r*}} |z|^{-p'}dxdy\Big)^{1/p'}\\ &\le (1/\pi)[2\pi/(2-p')]^{1/p'}(2r*)^{(2-p')/p'}\|g_m-g_n\|_p,\end{aligned}$$

where $p>2$ and $1/p+1/p'=1$, and this, together with $(2-p')/p'=1-2/p$ and (9.4), suffices to conclude that $f_n \rightrightarrows f$.

S t e p E. Owing to Lemma 13 we know that each f_n represents a regular Q_n-qc homeomorphism with fixpoint at ∞, where $Q_n \to Q$, Q being given by (9.2) with μ replaced by μ^*. Hence, by Step D and the definition of Q-quasiconformality, in order to conclude that $af+b$ is a homeomorphism of S_Q^* for some constants a and b, it remains to prove that f is one-to-one.

S t e p F. Since each f_n is one-to-one, we can consider $\check{f}_n$. At first we are going to prove that $\omega = \check{f}_{n\bar{w}}(w)$ is a solution of (7.1)

<u>with</u> $\Phi = \mathrm{id}$ <u>and</u> μ <u>replaced by</u> $-(\mu \circ \check{f}_n)\exp(-2i \arg \check{f}_{nw})$.

Owing to the fact that the complex dilatation of id is 0, we have, by (9.1),

$$(9.5) \quad \check{f}_{n\bar{w}} = -(\mu_n \circ \check{f}_n)\exp(-2i \arg \check{f}_{nw})\check{f}_{nw}.$$

On the other hand, arguing as in Step D, we have

$$|f_n(s) - s| \le (1/\pi)\iint_{\Delta^r} |z - s|^{-1} |g_n(z)|\,dxdy$$
$$\le (1/\pi)[2\pi/(2 - p')]^{1/p'}(2r^*)^{1-2/p} \|g_n\|_p,$$

where $p > 2$ and $1/p + 1/p' = 1$, and this, together with (9.3), yields the boundedness of $f_n - \mathrm{id}$ by a constant independent of n. Consequently there is an $\tilde{r}$, $0 < \tilde{r} < +\infty$, such that $\Delta^{\tilde{r}} \supset \bigcup \operatorname{spt} \check{f}_{n\bar{w}}$. Therefore $\check{f}_{n\bar{w}}$ is an L^1-function and, by Lemma 4, we get

$$(9.6) \quad \check{f}_n = \Phi + T\check{f}_{n\bar{w}},$$

Φ_n being holomorphic (in $\mathbb{C}$).

It is clear that $T\check{f}_{n\bar{w}}$ is bounded by a constant depending, in general, on n with $(T\check{f}_{n\bar{w}})(w) \to 0$ for $w \to \infty$. On the other hand, as we have already proved, $\check{f}_n - \mathrm{id}$ is bounded and, since $|\check{f}_n - \mathrm{id}| = |(Tg_n) \circ \check{f}_n|$, we have $|\check{f}_n(w) - w| \to 0$ as $w \to \infty$. Consequently, by (9.6), $\Phi_n - \mathrm{id}$ is bounded and $\Phi(w) - w \to 0$ as $w \to \infty$, so, since it is holomorphic, $\Phi_n - \mathrm{id} = 0$ for each n by the classical theorem of Liouville. Therefore (9.6) becomes $\check{f}_n = \mathrm{id} + T\check{f}_{n\bar{w}}$, whence, by (9.5) and Lemma 9, $\omega = \check{f}_{n\bar{w}}(w)$ is a solution of (7.1) with $\Phi = \Phi_n = \mathrm{id}$ and μ replaced by $-(\mu \circ \check{f}_n)\exp(-2i \arg \check{f}_{nw})$, as desired.

S t e p G. The conclusion of Step F together with the relation $\operatorname{spt} \check{f}_{n\bar{w}} \subset \Delta^{\tilde{r}}$ allow us to utilize (9.3) with g_n replaced by $\check{f}_{n\bar{w}}$ and r by $\tilde{r}$, i.e.

$$(9.7) \quad \|\check{f}_{n\bar{w}}\|_p \le q(\pi\tilde{r}^2)^{1/p}(1 - q\|S\|_p)$$

<u>with</u> p <u>and</u> ε <u>as in</u> (9.3). <u>We claim that there is a subsequence</u> $\check{f}_{n_k} \rightrightarrows \check{f}$, say. In order to conclude this by the classical theorem of Ascoli and Arzelà, we have to prove that $\check{f}_n$ are equicontinuous. To this end we consider the difference $\check{f}_n(w_1) - \check{f}_n(w_2)$ for $w_1, w_2 \in \Delta^{r^*} \supset \Delta^{\tilde{r}}$. By (9.6), the relation $\Phi_n = \mathrm{id}$, the definition of T, and Hölder's inequality, we have

$$|\check{f}_n(w_1) - \check{f}_n(w_2)| \le |w_1 - w_2| + (1/\pi)\iint_{\Delta^{\tilde{r}}} |(w-w_1)^{-1} - (w-w_2)^{-1}| \times |\check{f}_{n\bar{w}}(w)|\,dudv$$
$$\le |w_1 - w_2| + (1/\pi)\|\check{f}_{n\bar{w}}\|_p |w_1 - w_2| \times \{\iint_{\Delta^{\tilde{r}}} (|w-w_1||w-w_2|)^{-p'} dudv\}^{1/p'},$$

where $u = \operatorname{re} w$, $v = \operatorname{im} w$, and $1/p + 1/p' = 1$. Let $\tilde{r}^* > r^*$. For $|w_1 - w_2| \le \tilde{r}^* - r^*$ we have $\Delta(w_1) \cup \Delta(w_2) \subset \Delta^{\tilde{r}^*}$, where $\Delta(a) = \{w\colon |w-a| \le \frac{1}{2}|w_1 - w_2|\}$, and also $|w-w_1||w-w_2| > \frac{1}{3}|w-w_1|^2$ whenever $w \in \Delta^{\tilde{r}^*} \setminus \Delta(w_1) \setminus \Delta(w_2)$ (cf. Fig. 3, p. 15). Consequently, under the notation $\Delta(w_1, w_2) = \{w\colon \frac{1}{2}|w_1 - w_2| \le |w-w_1| \le 2\tilde{r}^*\}$, for $p > 2$ and $|w_1 - w_2| \le \min(2, \tilde{r}^* - r^*)$, we get

$$\iint_{\Delta^{\tilde{r}^*}} \frac{dudv}{(|w-w_1||w-w_2|)^{p'}} \le \frac{2^{p'}}{|w_1-w_2|^{p'}} \iint_{\Delta(w_1)} \frac{dudv}{|w-w_1|^{p'}}$$
$$+ \frac{2^{p'}}{|w_1-w_2|^{p'}} \iint_{\Delta(w_2)} \frac{dudv}{|w-w_2|^{p'}} + 3^{p'} \iint_{\Delta(w_1,w_2)} \frac{dudv}{|w-w_1|^{2p'}}$$
$$= 4\pi(\tfrac{1}{2}|w_1-w_2|)^{2-2p'}\{\frac{1}{2-p'} + \frac{3^{p'}}{4(p'-1)}[1 - \frac{(2\tilde{r}^*)^{2-2p'}}{(\frac{1}{2}|w_1-w_2|)^{2-2p'}}]\}$$
$$\le 4\pi[\frac{1}{2-p'} + \frac{3^{p'}}{4(p'-1)}](\tfrac{1}{2}|w_1-w_2|)^{2-2p'}.$$

Therefore, by (9.7), we conclude that

$$(9.8)\quad |\check{f}_n(w_1) - \check{f}_n(w_2)| \le |w_1 - w_2| + 2^{(2+p')/p'}[\frac{1}{2-p'} + \frac{3^{p'}}{4(p'-1)}]^{1/p'} \frac{q\tilde{r}^{2/p}}{1 - q\|S\|_p}(\tfrac{1}{2}|w_1-w_2|)^{(2-p')/p'}$$

provided that $q\|S\|_p < 1$, $2 < p < 2 + \varepsilon$, and $|w_1 - w_2| \le \min(2, \tilde{r}^* - r^*)$, and this, together with $(2-p')/p' = 1 - 2/p$, proves that $\check{f}_n$ are equicontinuous, as desired.

S t e p <u>H</u>. <u>We now claim that</u> $\check{f} \circ f = \mathrm{id}$. In fact, since $f_n \rightrightarrows f$ and $\check{f}_{n_k} \rightrightarrows \check{f}$, then for any $\eta > 0$ and $r^* > 0$ there is an integer $k_o > 0$ such that $k > k_o$ implies

$$\max\{|f_{n_k}(z) - f(z)|,\ |\check{f}_{n_k} \circ f(z) - \check{f} \circ f(z)|\} < \eta \quad \text{for } z \in \Delta^{r^*}.$$

On the other hand,

$$|\check{f}_{n_k}\circ f_{n_k}(z) - \check{f}_{n_k}\circ f(z)| \le M|f_{n_k}(z) - f(z)|^{1-2/p} \quad \text{for } z \in \Delta^{r^*},$$

where $2 < p < 2+\varepsilon$ and M is a constant independent of k, determined in (9.8); here we assume that $|f_{n_k}(z) - f(z)| \le \min(2, \tilde{r}^* - r^*)$. Therefore, taking $\eta < \min(2, \tilde{r}^* - r^*)$, we get

$$|\tilde{f}\circ f(z) - z| \le |\tilde{f}\circ f(z) - \check{f}_{n_k}\circ f(z)| + |\check{f}_{n_k}\circ f(z) - \check{f}_{n_k}\circ f_{n_k}(z)|$$
$$< \eta + M\eta^{1-2/p} \longrightarrow 0 \quad \text{as } \eta \longrightarrow 0+,$$

whence $\tilde{f}\circ f = \mathrm{id}$, as desired.

S t e p <u>I</u>. In Step E we have remarked what follows: in order to conclude that the mapping $af+b$ is a homeomorphism of the class S_Q^* for some constants a and b, <u>it remains to prove that</u> f <u>is one-to-one</u>. We have shown in Step H that $\tilde{f}\circ f = \mathrm{id}$. Consequently, if $f(z_1) = f(z_2)$, then $z_1 = z_2$ indeed.

S t e p <u>J</u>. Finally, <u>it remains to prove that</u> $af+b$ <u>is determined uniquely</u>. To this end consider another solution $w = \hat{a}\hat{f}(z) + \hat{b}$ of (4.2) with the same properties as $w = af(z)+b$ (cf. the proof of Lemma 11). Both $f'_{\bar z}$ and $\hat{f}'_{\bar z}$ are, by Lemma 3, locally L^2, so – by Lemma 4 – we have

(9.9) $\quad f = \mathrm{id} + Tf'_{\bar z}, \quad \hat{f} = \mathrm{id} + T\hat{f}'_{\bar z}.$

Therefore, by Lemma 9, $\omega = f'_{\bar z}(z)$ and $\omega = \hat{f}'_{\bar z}(z)$ are solutions of (7.1) with $\Phi = \mathrm{id}$. Consequently, by Lemma 10,

(9.10) $\quad \hat{f}'_{\bar z} = f'_{\bar z}$ a.e.

Relations (9.9) and (9.10) yield $f = f$ and the normalization shows that $a = a$ and $b = b$, whence $af+b = af+b$ indeed. This completes the whole proof.

<u>COROLLARY 2</u>. <u>Any</u> Q-<u>qc</u> <u>homeomorphism</u> $f\colon \operatorname{int}\Delta \longrightarrow \operatorname{int}\Delta$ <u>can be continued to a</u> Q-<u>qc</u> <u>homeomorphism of</u> $\mathbb{E}$ <u>onto</u> $\mathbb{E}$ <u>which transforms</u> $\operatorname{fr}\Delta$ <u>onto itself</u>.

<u>P r o o f</u>. Owing to the properties of homographies we may, without any loss of generality, assume that $f(0) = 0$ and $f(z_n) \longrightarrow 1$ for some sequence of $z_n \longrightarrow 1$. By Lemmas 1 and 3, f possesses distributional derivatives f'_z, $f'_{\bar z}$ which satisfy (9.2) with $\mu = f'_{\bar z}/f'_z$ a.e. Consider now the equation (4.2) with μ replaced by μ^* defined as follows: $\mu^*(z) = \mu(z)$ for $z \in \operatorname{int}\Delta$, and $\mu^*(z) = e^{4i\arg z}\,\overline{\mu(1/\bar z)}$ otherwise. By Theorem 2 there exists a unique solution $w = f^*(z)$ of this equation

which represents a homeomorphism of S_Q^* with Q given by (9.2). We claim that

(9.11) $f^*(z) = 1/\overline{f^*(1/\bar{z})}$ for $z \neq 0, \infty$.

In fact, consider the function $f^{**}: \mathbb{E} \to \mathbb{E}$, defined by

(9.12) $f^{**}(z) = 1/\overline{f^*(1/\bar{z})}$ for $z \neq 0, \infty$; $f^{**}(z) = z$ for $z = 0, \infty$.

Since $f^{**} \in S_Q^*$ and the corresponding complex dilatation $\mu^{**} = \mu^*$ a.e. in $\mathbb{C}$ so that $w = f^{**}(z)$ is a solution of (4.2) with μ replaced by μ^*, then $f^{**} = f^*$, and this, together with (9.12), yields (9.11). From (9.11) it follows that f^* transforms $\operatorname{fr}\Delta$ onto itself. Finally, denoting $f^*|\operatorname{int}\Delta$ by $\hat{f}$, we prove exactly in the same way as in Step J of the preceding proof, that $\hat{f} = f$.

We conclude this section by an analogue of Theorem 2 for $f \in S_Q$.

COROLLARY 3. *Suppose that* μ *is a measurable function defined in* Δ *and such that* $\|\mu\|_\infty < 1$. *Then there exists a unique solution* $w = f(z)$ *of* (4.2) *which represents a homeomorphism belonging to* S_Q, *where* Q *is given by* (9.2).

Proof. Consider the equation (4.2) with μ replaced by μ^* defined as in the proof of Corollary 2. By Theorem 2 there exists a unique solution of this equation which represents a homeomorphism of S_Q^* with Q given by (9.2). In order to complete the proof it is now sufficient to verify that $|f^*(z)| = 1$ for $|z| = 1$ or, stronger, that (9.11) holds, and this can be done exactly in the same way as in the proof of Corollary 2.

Finally, it is worth-while to remark that the method of Bojarski works also in the case where (4.2) is replaced with

$$w'_{\bar{z}} - \mu(z)w'_z - \mu^*(z)\bar{w}'_{\bar{z}} = a(z)w + a^*(z)\bar{w} + b(z),$$

where μ and μ^* are measurable functions with $\|\mu\|_\infty + \|\mu^*\|_\infty < 1$, while a, a^*, and b are locally L^p for some $p > 2$ (for details see Bojarski [2]; for further extensions cf. Bojarski and Iwaniec [1]).

10. Extension to multiply connected domains

For proving the theorem on existence and uniqueness of qc mappings with a preassumed measurable complex dilatation in the case of multiply connected domains we essentially follow Ławrynowicz [1]. Actually the first existence proof was given by Parter [1], but it did not yield any proof of uniqueness. A direct proof independent of the corresponding theorem for conformal mappings of multiply connected domains is due to Renelt [2].

Under a circular domain we understand a domain of the form

$$(10.1) \quad \mathbb{E} \setminus \bigcup_{k=1}^{n} \Delta_k^*,$$

where Δ_k^* are disjoint circular sets, i.e. closed discs, one point sets or the complement of an open disc. Under a canonical circular domain we understand a circular domain (10.1), where $\Delta_1^* = \{z: |z| \geq 1\}$ and $\Delta_2^* = \Delta^r$ in case $n \geq 2$. We formulate the problem so that the image domain is sought in the form of a canonical circular domain or a closed canonical circular domain. From the theory of conformal mappings it is well known (cf. e.g. Courant [2], pp. 38-39) that two multiply connected domains of the same connectivity are, in general, not conformally equivalent. We shall see that the same holds for qc mappings, when the complex dilatation is preassigned a.e. in the domain of definition. Of course we can make an unessential generalization by replacing canonical circular domains with a more general class of canonical domains.

Let n be, as before, a positive integer. The domain D of definition of the mapping to be found is supposed to be the closure of a domain bounded by n disjoint Jordan curves. It is worth-while to remark that the problem makes also sense for arbitrary closed domains and actually so it is treated by Ławrynowicz [1], assuming, however, a finite connectivity, but one has to understand boundary points of a domain and the convergence to them in the sense of the theory of prime ends due to Carathéodory [1] (cf. e.g. Collingwood and Lohwater [1]) and to generalize the notion of qcty correspondingly. Finally, similarly to the case of simply connected domains, one can reduce the problem for domains to the problem for closed domains and vice versa.

THEOREM 3. *Suppose that μ is a measurable function defined in D and such that $\|\mu\|_\infty < 1$. Then there exists a closed canonical circular domain D' of connectivity n and a solution $w = f(z)$ of (4.2) which represents a Q-qc homeomorphism of D onto D', determined uniquely apart from conformal mappings of D' onto itself, where Q is given by (9.2).*

Proof. Let D_* be an arbitrary closed simply connected domain which is different from $\mathbb{E}$ and includes D. Let further μ_* be a measurable function defined in D_* so that $\|\mu_*\|_\infty \le \|\mu\|_\infty$ and $\mu_*|D=\mu$ a.e. Let finally D'_* be an arbitrary simply connected closed domain different from $\mathbb{E}$.

In view of Corollary 3 and remarks at the beginning of Section 9, there exists a solution $w=f(z)$ of (4.2) with μ replaced by μ^*, which represents a Q-qc homeomorphism of D_* onto D'_*, where Q is given by (9.2). This mapping transforms, in particular, the closed domain D onto a certain closed domain $D^* \subset D'_*$ Q-quasiconformally. Applying now the theorem on conformal mapping of multiply connected domains onto canonical circular domains (cf. e.g. Courant [2], pp. 169-178) and the classical theorem on homeomorphic extension of conformal mapping of domains bounded by n disjoint Jordan curves (cf. e.g. Behnke and Sommer [1], p. 371), we see that D^* can be conformally mapped onto a closed canonical circular domain D' of connectivity n. However, it can be easily seen from the formula (9.1) that a Q-qc mapping with the complex dilatation μ composed externally with a conformal mapping, gives again a Q-qc mapping with the complex dilatation μ. Thus, there exists a circular canonical closed domain D' of connectivity n and a Q-qc homeomorphism of D onto D' with the complex dilatation μ. Consequently, this homeomorphism is a desired solution of (4.2).

We still have to prove that the described mapping of D onto D' is determined uniquely apart from conformal mappings of D' onto itself. Let us suppose that there exist two solutions $w=f(z)$ and $w=\hat{f}(z)$ which fulfil the conditions of our theorem. By the above consideration, $f=\Phi_1 \circ f^*$ and $\hat{f}=\Phi_2 \circ \hat{f}^*$, where Φ_1 and Φ_2 are conformal and (cf. the proof of Lemma 11)

(10.2) $f^* = \mathrm{id}_D + Tf^{*\prime}_{\bar{z}}, \quad \hat{f}^* = \mathrm{id}_D + T\hat{f}^{*\prime}_{\bar{z}},$

T being defined in (4.5). Therefore, by Lemma 9, $\omega = f^{*\prime}_{\bar{z}}(z)$ and $\omega = \hat{f}^{*\prime}_{\bar{z}}(z)$ are solutions of (7.1) with $\Phi = \mathrm{id}_D$. Consequently, if we continue μ by the value 0 to the entire plane, we obtain, by Lemma 10,

(10.3) $\hat{f}^{*\prime}_{\bar{z}} = f^{*\prime}_{\bar{z}}$ a.e.

Relations (10.2) and (10.3) yield $\hat{f}^* = f^*$. Thus we conclude that $\hat{f} = \Phi_2 \circ \check{\Phi}_1 \circ f$, and this suffices to arrive at the desired result.

It is worth-while to remark that <u>the conclusion of Theorem 3 remains still valid for infinitely connected domains</u> D' provided D' is a <u>parallel slit domain</u>, i.e. the closed plane $\mathbb{E}$ slit along a collection of parallel straight-line segments which may reduce to one point sets (cf. e.g. Golusin [2], pp. 178-183).

Now we turn our attention to the particular case $n=2$. A function f is said to be of the class $S_Q^{r,R}$ if it maps A_r onto A_R Q-quasiconformally with $f(1)=1$; in the degenerate case $r=0$ we assume, in addition, that $f(0)=0$. Thus $S_Q^{0,0}=S_Q$. We will see later (Corollary 16) that the class $S_Q^{r,R}$ is nonempty iff $r^Q \leq R \leq r^{1/Q}$ (we shall always assume that $r<1$).

An immediate consequence of Theorem 3 and the formula (9.1) is the following

COROLLARY 4. Suppose that μ is a measurable function defined in a nondegenerate annulus A_r and such that $\|\mu\|_\infty < 1$. Then there exists exactly one number R, $0<R<1$, and a unique solution $w=f(z)$ of (4.2) which represents a homeomorphism belonging to $S_Q^{r,R}$, where Q is given by (9.2).

11. Some equivalent characterizations of quasiconformal mappings

We can give now the so-called analytic characterization of Q-qc mappings (in two alternative formulations). It is partly due to Mori [1], Bers [1, 3], Gehring and Lehto [1], Pfluger [2], and Ahlfors and Bers [1].

THEOREM 4. A mapping $f: D \to D'$, where D and D' are domains in $\mathbb{E}$, is Q-qc iff it is a sense-preserving homeomorphism which possesses distributional L^2-derivatives f'_z and $f'_{\bar z}$, satisfying (4.1).

Proof. Suppose first that D is of finite connectivity and f is Q-qc. Then there is a sequence of regular Q-qc mappings $f_n: D_n \to D'_n$ such that $f_n \rightrightarrows f$ and for a.e. z for which there exist partial derivatives f_z, $f_{\bar z}$ we have $\mu_n(z) \to \mu(z)$, where $\mu_n = f_{n\bar z}/f_{nz}$ and $\mu = f_{\bar z}/f_z$. Owing to the theorem on conformal mapping of multiply connected domains (in particular, simply connected) onto circular canonical domains as well as to the Riemann mapping theorem for the cases of the punctured and closed planes we may fix a suitable number of parameters in order to assure that, in Theorem 3, each of the equations (4.2) and (4.2) with μ replaced by μ_n, $n=1,2,\ldots$, possesses exactly one solution which represents a Q-qc homeomorphism between the corresponding domains; denote them by $\hat f$ and $\hat f_n$, $n=1,2,\ldots$, respectively. Since we already know that for each n the corresponding f_n fulfils exactly the same conditions as $\hat f_n$, we conclude that $\hat f_n = f_n$ and hence also $\hat f = f$. Thus f is a sense-preserving homeomorphism. Therefore, by

Lemmas 1 and 3, f satisfies (2.2) and possesses distributional L^2-derivatives $f'_z = f_z$ and $f'_{\bar z} = f_{\bar z}$ a.e., whence also (4.1) holds.

Suppose now that f is a sense-preserving homeomorphism which possesses distributional L^2-derivatives f'_z and $f'_{\bar z}$, satisfying (4.1), while D is, as before, of finite connectivity. Continuing $\mu = f'_{\bar z}/f'_z$ to the entire plane by the value 0 and arguing exactly in the same way as in the proof of Theorem 2, Steps A-E, with the only change that we have to compose externally the constructed regular Q-qc mappings f_n with suitable conformal mappings Φ_n, we prove that there is a sequence of regular Q-qc mappings $\Phi_n \circ f_n$: $D \to D'$ such that $\Phi_n \circ f_n \rightrightarrows f$ and for a.e. z for which there exist partial derivatives $f_z(z)$, $f_{\bar z}(z)$ we have $f_{n\bar z}(z)/f_{nz}(z) \to f_{\bar z}(z)/f_z(z)$, as desired.

The case of infinitely connected D can be treated exactly in the same way as the preceding case provided we take into account the remark after the proof of Theorem 3.

A slight modification of the preceding argument or a direct application of Theorem 4 together with Lemma 3 yields the following alternative formulation of this theorem:

COROLLARY 5. *A mapping* $f: D \to D'$, *where* D *and* D′ *are domains in* $\mathbb{E}$, *is* Q-*qc iff it is an ACL sense-preserving homeomorphism and* (2.2) *holds*.

We conclude this chapter by a survey of other characterizations of the class of Q-qc mappings between two fixed plane domains D and D′. In this survey we essentially follow Gehring [2].

Let Σ be a configuration consisting of a domain D bounded by m Jordan curves, together with n boundary points and p interior points distinguished. Then there are exactly four cases where Σ has precisely one conformal invariant $\mathfrak{I}(\Sigma)$ apart from one-to-one mappings of the two-point-compactification of $\mathbb{R}$ (cf. Ahlfors [1], p. 88, or [7], pp. 70–71):

1° $m=2$, $n=0$, $p=0$: $\mathfrak{I}(D) = \operatorname{mod} D$,

2° $m=1$, $n=4$, $p=0$: $\mathfrak{I}(D; z_1, z_2, z_3, z_4) = \operatorname{mod}(D, \Gamma, \Gamma^*)$,

3° $m=1$, $n=2$, $p=1$: $\mathfrak{I}(D; z_1, z_2; z) = \omega(z, \Gamma, D)$,

4° $m=1$, $n=0$, $p=2$: $\mathfrak{I}(D;; z_1, z_2) = h(z_1, z_2, D)$,

where the symbols standing on the right-hand sides of the above formulae for $\mathfrak{I}$ denote:

1° the modulus of the doubly connected domain D,

2° the modulus of the quadrilateral D with distinguished boundary arcs $\Gamma = \widehat{z_1 z_2}$ and $\Gamma^* = \widehat{z_3 z_4}$,

3° the harmonic measure of the boundary arc $\Gamma = \widehat{z_1 z_2}$ of D at z, where D is a Jordan domain,

4° the hyperbolic distance between z_1 and z_2 w.r.t. the bounded simply connected domain D (for the definitions of these familiar notions cf. e.g. Ahlfors [5], pp. 13-14 and 21, and Krzyż [2], pp. 114 and 118).

A generalization of the above conformal invariants is given by the notion of the modulus $\operatorname{mod}\{\Gamma\}$ of a family $\{\Gamma\}$ of Jordan arcs or curves. Let $\operatorname{adm}\{\Gamma\}$ denote the family of functions ρ which are non-negative and Borel measurable in $\mathbb{E}$ and for which the integral of ρ over Γ with respect to the arc length is not less than 1 for all locally rectifiable Γ in $\{\Gamma\}$. We recall that a Jordan arc which does not contain ∞ is said to be *locally rectifiable* if every of its closed subarcs is rectifiable. A Jordan arc which contains ∞ is said to be *locally rectifiable* if there exists a subarc of it which contains ∞ and a reflection that maps this subarc onto a locally rectifiable arc which does not contain ∞. A locally rectifiable Jordan curve is defined similarly. Now we are ready to define the *modulus* of $\{\Gamma\}$ as

$$\operatorname{mod}\{\Gamma\} = \inf_{\rho} \iint_{\mathbb{C}} \rho^2\, dx dy, \quad \rho \in \operatorname{adm}\{\Gamma\},$$

and the *extremal length* $\lambda\{\Gamma\} = 1/\operatorname{mod}\{\Gamma\}$. The notion of extremal length is due to Ahlfors and Beurling [1, 2]. Conformal invariants appearing in 1°-4° can be expressed in terms of moduli of certain arc or curve families connected with a given configuration.

We can now formulate the so-called *geometric characterizations* of Q-qc mappings.

A sense-preserving homeomorphism $f: D \to D'$ *is* Q-*qc iff one of the following (equivalent) conditions holds:*

(i) $\lambda(f\{\Gamma\}) \leq Q\lambda\{\Gamma\}$ *or, equivalently,* $\lambda(f\{\Gamma\}) \geq Q^{-1}\lambda\{\Gamma\}$ *for any family* $\{\Gamma\}$ *of Jordan arcs or Jordan curves* Γ *in* D (Väisälä [1]).

(ii) $\operatorname{mod} f[D_0] \leq Q \operatorname{mod} D_0$ *or, equivalently,* $\operatorname{mod} f[D_0] \geq Q^{-1} \operatorname{mod} D_0$ *for any doubly connected domain* D_0 *with* $\operatorname{cl} D_0$ D (Gehring and Väisälä [1], Reich [1]).

(ii′) $\operatorname{mod} f[D_0] \geq \frac{1}{2}(Q + 1/Q)^{-1} \operatorname{mod} D_0$ for any annulus D_0 with $\operatorname{cl} D_0 \subset D$ (ibid.).

(iii) $\operatorname{mod} f[\mathbb{D}] \leq Q \operatorname{mod} \mathbb{D}$ *or, equivalently,* $\operatorname{mod} f[\mathbb{D}] \geq Q^{-1} \operatorname{mod} \mathbb{D}$ *for any quadrilateral* $\mathbb{D} = (D_0, \Gamma, \Gamma^*)$ *with* $\operatorname{cl} D_0 \subset D$ (Ahlfors [2], cf. also Pfluger [1]). Historically, this is the first definition of Q-qc mappings in the contemporary sense.

(iii′) The same condition restricted to $\mathbb{D}$ being rectangles (Gehring and Väisälä [1]).

(iii″) The same condition restricted to $\mathbb{D}$ being quadrilaterals with $\operatorname{mod}\mathbb{D} = 1$ (Kelingos [1]).

(iii‴) $\operatorname{mod} f[\mathbb{D}] \le \frac{1}{2}(Q + 1/Q)\operatorname{mod}\mathbb{D}$ or, equivalently, $\operatorname{mod} f[\mathbb{D}] \ge 2(Q + 1/Q)^{-1}\operatorname{mod}\mathbb{D}$ for any rectangle $\mathbb{D} = (D_0, \Gamma, \Gamma^*)$ with $\operatorname{cl} D_0 \subset D$ and Γ parallel to a fixed straight line (Gehring and Väisälä [1]).

(iv) $\nu \circ \sin\frac{1}{2}\pi\omega(f(z), f[\Gamma], f[D_0]) \le Q\,\nu \circ \sin\frac{1}{2}\pi\omega(z, \Gamma, D_0)$ or, equivalently, the left-hand side of this inequality is not less than $Q^{-1} \times \nu \circ \sin\frac{1}{2}\pi\omega(z, \Gamma, D_0)$ for any Jordan domain D_0 with $\operatorname{cl} D_0 \subset D$, any arc $\Gamma \subset \partial D_0$ and any $z \in D_0$, where $\nu(r)$ is the modulus of $\Lambda \setminus [0; r]$ or, equivalently, $\nu(r) = \frac{1}{4}K((1 - r^2)^{\frac{1}{2}})/K(r)$, K being the complete elliptic normal integral of the first kind (Kelingos [1], Hersch and Pfluger [1]).

(v) $\nu \circ \exp(-2h(f(z_1), f(z_2), f[D_0])) \le Q\,\nu \circ \exp(-2h(z_1, z_2, D_0))$ or, equivalently, the left-hand side of this inequality is not less than $Q^{-1}\,\nu \circ \exp(-2h(z_1, z_2, D_0))$ for any bounded simply connected domain D_0 with $\operatorname{cl} D_0 \subset D$ and any pair of points $z_1, z_2 \in D_0$ (ibid., Hersch [1]).

We turn next our attention to characterizations of Q-qc mappings in terms of distortion of circular neighbourhoods and angles, called metric characterizations. Historically, the problem of distortion for circular neighbourhoods inspired Lavrentieff [1, 2] to define a class of Q-qc mappings. These results were next generalized by Pesin [1] and Gehring [1]. The problem of distortion of angles by qc mappings was independently investigated by Agard and Gehring [1] and Taari [1]. We begin with the definitions of circular dilatation (according to Lavrentieff [2]) and of an angle and its inner measure (according to Agard and Gehring [1]). We formulate the corresponding characterization as it is done in Gehring [1] and Agard and Gehring [1], respectively.

If f is a homeomorphism of $D \subset \mathbb{C}$ into $\mathbb{C}$, then the circular dilatation of f at $z \in D$, denoted by $H(z)$ or $H_f(z)$, is defined as

$$H_f(z) = \limsup_{t \to 0+} \frac{\max_\alpha |f(z + te^{i\alpha}) - f(z)|}{\min_\alpha |f(z + te^{i\alpha}) - f(z)|}, \quad \alpha \text{ real}.$$

We generalize this definition to the cases where D or $f[D]$ contains ∞, putting $H_f(\infty) = H_{f^*}(0)$ and $H_f(z_0) = H_{f^{**}}(z_0)$ for $z = \check{f}(\infty)$, where f^* and f^{**} are defined by $f^*(z) = f(1/z)$ and $f^{**}(z) = 1/f(z)$, respectively.

I. Basic concepts and theorems

We say that two Jordan arcs Γ_1 and Γ_2 form an angle at a point z_0, if they meet only at z_0 which is their common end point. Then we define the inner measure $A(\Gamma_1,\Gamma_2)$ of this angle as

$$A(\Gamma_1,\Gamma_2) = \liminf_{z_1,z_2 \to z_0} 2 \arcsin(|z_1 - z_2|/a(z_0,z_1,z_2)),$$

where $z_1 \in \Gamma_1$, $z_2 \in \Gamma_2$, $a(z_0,z_1,z_2) = |z_1 - z_0| + |z_2 - z_0|$ for $z_0 \neq \infty$ and $a(z_0,z_1,z_2) = |z_1| + |z_2|$ for $z_0 = \infty$.

A sense-preserving homeomorphism $f: D \to D'$ is Q-qc iff one of the following (equivalent) conditions holds:

(vi) H is bounded everywhere and bounded by Q a.e.

(vii) Firstly, $A(f[\Gamma_1], f[\Gamma_2]) > 0$ for any $z_0 \in D$ and any pair of Jordan arcs $\Gamma_1, \Gamma_2 \subset D$, which form an angle at z_0, and, secondly, $A(f[\Gamma_1], f[\Gamma_2]) \geq Q^{-1} A(\Gamma_1,\Gamma_2)$ or, equivalently, $A(f[\Gamma_1], f[\Gamma_2]) \leq Q \times A(\Gamma_1,\Gamma_2)$ for a.e. $z_0 \in D$ and every pair of Jordan arcs $\Gamma_1, \Gamma_2 \subset D$, which form an angle at z_0.

We mention that there are also various equivalent characterizations of the class of all qc mappings (with Q not specified), not only such that can formally be derived from the above characterizations, but also essentially different. This topic is treated in detail in Gehring [2].

Finally, before we begin to deal with the problem of parametrization for qc mappings, we remark that it is closely connected with investigations of qc mappings near to the identity mapping which will not be studied in these lecture notes. We only mention the fundamental fact discovered by Ahlfors [4] that every qc mapping can be expressed as the composition of a finite number of qc mappings – Q_j-qc, $j = 1,\ldots,n$, say, so that the $\max Q_j$ be arbitrarily close to 1. These mappings were also treated by Belinskiĭ [1], Cheng Bao-long [1], Schiffer [2], Krzyż and Ławrynowicz [1], Ławrynowicz [3], and Reich and Strebel [3]. As a tool they appear in Section 13 below.

II. THE PARAMETRICAL METHODS

12. Homotopical deformations of quasiconformal mappings and a lemma on asymptotic behaviour

In this chapter, for the sake of convenience, we shall often use the notation $w=f(s)$ instead of $w=f(z)$ in order to reserve the letter z for a variable of integration.

From the considerations of Sections 9 and 10 follows that when investigating qc mappings of simply and doubly connected domains we may, in general, confine ourselves to the classes $S_Q^{r,R}$. One of the most powerful tools when studying their extremal properties is given by the so-called parametrization theorems. For a dense subclass of $S_Q = S_Q^{0,0}$ basic theorems on parametrization were obtained by Shah Tao-shing [1] by using arbitrary sufficiently regular homotopies determined by the complex dilatations $\mu(\ ,t) = f_{\bar{s}}(\ ,t)/f_s(\ ,t)$, where $0 \le t \le T = \log Q$ and $f(\ ,T) = f \in S_Q$, which join f to id_Δ (cf. Corollary 3). A generalization for the case of $S_Q^{r,R}$ was obtained by Ławrynowicz [1].

The proofs of some of these theorems were next replaced with much simpler ones and the results extended to the whole classes $S_Q^{r,R}$ by Kruškal [1], and Gehring and Reich [1] in the case of S_Q, while by Ławrynowicz [2] in the general case of $S_Q^{r,R}$, however – even in the case of S_Q – for very special cases of homotopies only. These homotopies are determined by the equations

$$\mu(s,t) = (t/T)\mu(s), \tag{12.1}$$

$$\mu(s,t) = \begin{cases} \exp[i\arg\mu(s)]\tanh\{\frac{t}{T}\operatorname{arc\,tanh}|\mu(s)|\} & \text{for } s \neq 0, \\ 0 \text{ for } s=0. \end{cases} \tag{12.2}$$

Homotopy determined by (12.1) is due to Kruškal [1], while (12.2) to Gehring and Reich [1]. Unfortunately, these special homotopies cannot be, in general, used when studying subclasses of $S_Q^{r,R}$ since the fact that the function f belongs to a class S_* in question does not necessarily imply that $f(\ ,t)$ belongs to S_* for $0<t<T$, and simple

counterexamples can be given (cf. Gehring [3], and Reich and Strebel [1]).

We are going to give here eight theorems which solve the problem completely in the general case. The first of them, discussed in the next section, was given in an extremely concise formulation by Ahlfors and Bers [1], Lemma 21 and Theorem 10, and then by Ahlfors [5], p. 105, but even in the case of the homotopy determined by (12.2) some comments, given by Gehring and Reich [1], pp. 5-6, are needed. The other theorems (Theorems 6-12 in these lecture notes) were obtained by Ławrynowicz [5], where also the above mentioned theorem of Ahlfors and Bers was commented on.

In order to prove the theorem of Ahlfors and Bers we need two lemmas and in our presentation we essentially follow Ahlfors [5], pp. 100-106.

LEMMA 15. *Suppose that* μ *is a measurable function defined in* $\mathbb{C}$ *and such that* $\|\mu\|_\infty < 1$, *while* f *is the corresponding* S_Q^*-*solution of* (4.2); *cf.* Theorem 2. *Then*

$$(12.3)\qquad \|f'_z - 1\|_p \longrightarrow 0 \quad \text{as} \quad \|\mu\|_\infty \longrightarrow 0+ \quad \text{for} \quad p \geq 1.$$

Proof. In analogy to the proof of Theorem 2 (Step A) let us write $f = f_2^* \circ f_1$, $\check{f}_1 \in S_{Q_1}^*$, $f_2^* \in S_{Q_2}^*$, where the complex dilatations of f_1 and f_2^* are μ_1 and

$$\mu_2^* = \Big(\frac{\mu - \mu_1}{1 - \mu\bar{\mu}_1} \circ \check{f}_1\Big)\exp(-2i\arg \check{f}'_{1s}),$$

respectively. In contrast to the quoted proof, now it is more convenient to have the dilatation μ_1 (of f_1) with compact support. Thus, if this be the case, let $\mu = \mu_1 + \mu_2$, and let $\mu_2 | \operatorname{spt}\mu_1 = 0$. Clearly $\mu_2^* | \operatorname{spt}\mu_1 = 0$ as well. Let us specialize μ_1 so that it vanishes outside of Δ, i.e. we take there $\mu_2(s) = \mu(s)$, and so that $\mu_1(s) = \mu(s)$ inside of Δ. Therefore, by Lemmas 1, 2, and 14, especially formula (8.1), we have

$$f'_z = (f^{*\prime}_{2z} \circ f_1)f'_{1z} + (f^{*\prime}_{2\bar{z}} \circ f_1)\bar{f}'_{1z} = (f^{*\prime}_{2z} \circ f_1)f'_{1z}.$$

Hence

$$\|f'_z - 1\|_p \leq I_2 + I_1,$$

where

$$I_1 = \|f'_{1z} - 1\|_p, \quad I_2 = \|[(f^{*\prime}_{2z} - 1) \circ f_1]f'_{1z}\|_p.$$

Taking into account I_1, we apply Lemmas 9 and 10, by which

(12.4) $f_1 = \Phi + Tf'_{1\bar{z}}$,

where Φ is meromorphic in $\mathbb{E}$ and has the only (simple) pole at ∞, $D = \mathbb{C}$, and $\omega = f'_{1\bar{z}}(z)$ is an L^p-solution of the Tricomi singular integral equation (7.1). By Lemma 11 we can take here p arbitrarily large provided $\|\mu\|_\infty$ is sufficiently small. Thus, since $f'_{1\bar{z}} = \mu_1 f'_{1z}$, (7.1) becomes

(12.5) $f'_{1z} - S\mu_1 f'_{1z} = \Phi'$.

Next, since Φ has the only (simple) pole at ∞, there are constants a and b such that on substituting $af^*_1 + b$ for f_1 (12.5) becomes $f^{*\prime}_{1z} - 1 = S\mu^{*\prime}_{1z}$. Consequently, for any $p \geq 1$, we obtain, by Lemma 8,

$$\|f^{*\prime}_{1z} - 1\|_p \leq \|S\|_p \|\mu\|_p \|f^*_{1z} - 1\|_p + \|S\|_p \|\mu\|_p, \quad \|S\|_p < +\infty,$$

whence, by the relation $\|\mu_1\|_\infty \leq \|\mu\|_\infty$,

$$\|f'_{1z} - 1\|_p \leq \frac{\|S\|_p \|\mu_1\|_p}{1 - \|S\|_p \|\mu_1\|_p} \leq \frac{\|S\|_p \|\mu\|_\infty |\operatorname{spt}\mu_1|^{1/p}}{1 - \|S\|_p \|\mu\|_\infty |\operatorname{spt}\mu_1|^{1/p}} \longrightarrow 0$$
$$\text{as } \|\mu\|_\infty \longrightarrow 0+.$$

Finally, we observe that the relation $f_1 \in S^*_{Q_1}$ implies $a = 1/f^*_1(1)$ and hence, by (12.4), $a \longrightarrow 1$ as $\|\mu\|_\infty \longrightarrow 0+$. Thus

(12.6) $I_1 \longrightarrow 0$ as $\|\mu\|_\infty \longrightarrow 0+$.

We proceed now to consider I_2. Changing the variables and applying (1.5), we have

$$I_2^p \leq \iint_{f_1[\Delta]} |f^{*\prime}_{2z} - 1|^p \, |f'_{1z}|^p (1/|J_{f_1}|)\, dxdy$$
$$\leq (1 - \|\mu_1\|^2_\infty)^{-1} \iint_{f_1[\Delta]} |f^{*\prime}_{2z} - 1|^p \, |f'_{1z}|^{p-2}\, dxdy.$$

By the relation $\|\mu_1\|_\infty < \|\mu\|_\infty$ and the Schwarz inequality we infer that

$$I_2^p \leq (1 - \|\mu_1\|^2_\infty)^{-1} \Big(\iint_{f_1[\Delta]} |f^{*\prime}_{2z} - 1|^{2p}\, dxdy \Big)^{\frac{1}{2}}$$

$$\times \Big[\iint_\Delta |f'_{1z}|^{2p-4} (1 - |f'_{1\bar{z}}/f'_{1z}|^2)\, dxdy \Big]^{\frac{1}{2}}.$$

In order to conclude

(12.7) $\mathfrak{l}_2 \to 0$ as $\|\mu\|_\infty \to 0+$

it suffices now to verify that

(12.8) $$\iint_{f_1[\Delta]} |f^{*\prime}_{2z} - 1|^{2p}\, dxdy \to \text{ as } \|\mu\|_\infty \to 0+.$$

To this end we apply, as before, Lemmas 9, 10, and 11, showing that to prove (12.8) it is sufficient to verify that

(12.9) $$\mathfrak{l}_3^{2p} = \iint_{f_1[\Delta]} |f^{*\prime}_{3z} - 1|^{2p}\, dxdy \to 0 \text{ as } \|\mu\|_\infty \to 0+,$$

where $f_3^* = \mathrm{id} + Tf^{*\prime}_{3z}$. Next, we consider the mapping f_3, given by

(12.10) $f_3(z) = 1/\,\overline{f_3^*(1/\bar{z})}$, $z \in \mathbb{E}$.

Then, applying our conclusion (12.6) concerning f_1^* to f_3 and the holomorphy of f_3 for large $|z|$, we have e.g.

(12.11) $$\iint_{f_1[\Delta]} |f'_{3z}(z) - 1|^{4p}\, dxdy \to 0 \text{ as } \|\mu\|_\infty \to 0+.$$

On the other hand, changing the variables in $\mathfrak{l}_3$, we have

$$\mathfrak{l}_3^{2p} = \iint_{\Delta^*} |[z/f_3(z)]^2 f'_{3z}(z) - 1|^{2p} (1/|z|)^4\, dxdy$$

$$= \iint_{\Delta^*} |[z/f_3(z)]^2 [f'_{3z}(z) - 1] + [z/f_3(z)]^2 - 1|^{2p} (1/|z|)^4\, dxdy,$$

$$\Delta^* = \{z\colon\ 1/z \in f_1[\Delta]\},$$

whence, by the Minkowski and Schwarz inequalities,

$$\mathfrak{l}_3 \le \Big[\iint_{\Delta^*} (|z|^{8p-1}/|f_3(z)|^{8p}\, dxdy\Big]^{1/4p} \Big[\iint_{\Delta^*} |f'_{3z}(z) - 1|^{4p}\, dxdy\Big]^{1/4p}$$

$$+ \Big[\iint_{\Delta^*} |z^2/f_3^2(z) - 1|^{2p} (1/|z|)^4\, dxdy\Big]^{1/2p}.$$

We now check as in the proof of Theorem 2 (Step D) that $f_3 \rightrightarrows \mathrm{id}$ as $\|\mu\|_\infty \to 0+$. Therefore, by (12.11), we conclude (12.9), and thus also (12.8) and (12.7). This, together with (12.6), yields the desired relation (12.3).

13. Parametrization for mappings in the unit disc close to the identity mapping

We also need the following

LEMMA 16. *Let* φ *and* $\varepsilon(\ ,\tau)$, $0<\tau<T^*$, *be bounded measurable functions defined in* Δ, *where* $\|\varepsilon(\ ,\tau)\|_\infty \le \text{const}$ *and* $\varepsilon(\ ,\tau)\to 0$ *as* $\tau\to 0+$ *a.e. in* Δ. *Then there exists the limit of* $(1/\tau)[F(\ ,\tau)-\text{id}_\Delta]$ *as* $\tau\to 0+$ *in any Banach space* $B_p(\Delta)$, $p>2$, *with the norm*

$$\|f\|_{B_p(\Delta)} = \sup_{s_1,s_2\in\Delta} \frac{|f(s_1)-f(s_2)|}{|s_1-s_2|^{1-2/p}} + \|f'_z\|_p + \|f'_{\bar z}\|_p,$$

where $F(\ ,\tau)$ *belong to* $S_{Q(\tau)}$ *with* Q *given by* (9.2) *and are generated by the complex dilatations* $\mu(\ ,\tau)=\tau[\varphi+\varepsilon(\ ,\tau)]$; *cf*. *Corollary* 3. *This limit is given by the formula*

$$(13.1)\quad \lim_{\tau\to 0+}\frac{F(s,\tau)-s}{\tau} = \frac{s(1-s)}{\pi}\iint_\Delta\Big[\frac{\varphi(z)}{z(1-z)(s-z)}+\frac{\overline{\varphi(z)}}{\bar z(1-\bar z)(1-s\bar z)}\Big]dxdy,$$

where $x=\text{re}\,z$, $y=\text{im}\,z$, *and* $\iint_D = \lim_{\eta\to 0+}\iint_{D\setminus\Delta^\eta(s)}$.

Proof. For greater clarity the proof is divided into five steps.

Step A. By Lemmas 1, 3, and 4, for $|s|<1$ and $0<t<T^*$ we have

$$(13.2)\quad F(s,t) = \Phi(s,t) + T F'_{\bar s}(s,t),$$

where T is the operator defined in (4.5), the functions $\Phi(\ ,\tau)$ are holomorphic, and

$$(13.3)\quad T F'_{\bar s}(s,t) = -(1/\pi)\iint_\Delta F'_{\bar z}(z,t)(z-s)^{-1}\,dxdy.$$

We are going to show that

$$(13.4)\quad \Phi(s,t) = (1/2\pi i)\int_{\partial\Delta} F(z,t)(z-s)^{-1}\,dxdy.$$

To this end we apply the classical Cauchy integral formula: by (13.2), we have

$$(13.5)\quad \Phi(s,t) = (1/2\pi i)\int_{\partial\Delta^r}[F(z,t)-T F'_{\bar s}(z,t)](z-s)^{-1}\,dz,\quad |s|<r<1.$$

By (13.3) and Hölder's inequality, we get

$$|T F'_{\bar{s}}(s,t)| \le (1/2\pi)\, \|F'_{\bar{s}}(\ ,t)\|_p \Big(\iint_{\Delta} |z-s|^{-p'} dxdy\Big)^{1/p'}$$
$$\le (1/\pi)\, \|F'_{\bar{s}}(\ ,t)\|_p \Big(\iint_{\Delta^2} |z|^{-p'} dxdy\Big)^{1/p'}$$
$$\le (1/\pi)\,[2\pi/(2-p')]^{1/p'}\, 2^{(2-p')/p'}\, \|F'_{\bar{s}}(\ ,t)\|_p ,$$

where $p > 2$ and $1/p + 1/p' = 1$. On the other hand $T f'_{\bar{s}}(\ ,t)$ is holomorphic outside of Δ and vanishes at infinity, so, since $F(\ ,t)$ is continuous, we obtain (13.4) from (13.5) as the limit of (13.5) as $r \to 1-$.

S t e p B. Write now

$$(13.6)\qquad \Phi(s,t) = \frac{1}{2\pi i}\int_{\partial\Delta} \frac{F(z,t)}{z}\,dz + \frac{s}{2\pi i}\int_{\partial\Delta} \frac{F(z,t)}{z^2}\,dz + s^2\,\Psi(s,t)$$

(it is clear that both integrals on the right-hand side exist). <u>It is natural to divide the further proof into the following three steps</u>: C – to express the third addend on the right-hand side of (13.6) in terms of F as a double integral, D – to express the first two addends in terms of F as double integrals using the normalization conditions $F(0,t) = 0$ and $F(1,t) = 1$, $0 < t < T^*$, E – to prove the existence and calculate the limit (13.1) in $B_p(\Delta)$.

S t e p C. <u>We proceed to express the third addend on the right-hand side of</u> (13.6) <u>in terms of</u> F <u>as a double integral</u>. By (13.6), we have

$$2\pi i\, s^2\,\Psi(s,t) = \int_{\partial\Delta} F(z,t)\Big(\frac{1}{z-s} - \frac{1}{z} - \frac{s}{z^2}\Big)dz = s^2 \int_{\partial\Delta} \frac{F(z,t)}{z-s}\,\frac{dz}{z^2}.$$

Since $F(\ ,t) \in S_{Q(t)}$, we also have $F(s,t)\overline{F(s,t)} = 1$ for $s\bar{s} = 1$. Hence it follows that Ψ can be expressed by three other equivalent formulae

$$2\pi i\,\Psi(s,t) = \int_{\partial\Delta} \frac{1/\overline{F(z,t)}}{z-s}\,\frac{dz}{z^2} = -\int_{\partial\Delta} \frac{F(z,t)}{1-s\bar{z}}\,\bar{z}\,d\bar{z} = -\int_{\partial\Delta} \frac{1/\overline{F(z,t)}}{1-s\bar{z}}\,\bar{z}\,d\bar{z}.$$

We wish to utilize now one of the Green's formulae (3.2), but we have to prove it under weaker assumptions. Since we want to take in this formula $D^* = \operatorname{int}\Delta$, it is clear that we have to choose between two last formulae for Ψ, and comparing (4.6) with (4.9) we naturally decide to use the latter one:

(13.7) $\Psi(s,t) = -\frac{1}{2\pi i}\int_{\partial\Delta} \frac{1/\overline{F(z,t)}}{1-s\bar{z}}\bar{z}\,d\bar{z}.$

Then it is natural to expect that

(13.8) $\Psi(s,t) = \frac{1}{\pi}\iint_{\Delta} \frac{\partial}{\partial z}\frac{1/\overline{F(z,t)}}{1-s\bar{z}}\bar{z}\,dxdy.$

In order to prove (13.8), we apply again Lemma 4 with $D = \operatorname{int}\Delta \setminus \{0\}$ and

(13.9) $G(z,t) = z/F(z,t)$

instead of $f(w)$. This is justified since, by Theorem 1, $F'_{\bar{z}}$ is an L^p-function with arbitrary $p>2$ provided t, $t>0$, and thus also $Q(t)-1$ is sufficiently small; hence, by Minkowski's and Hölder's inequalities, for any annulus Δ_δ, $0<\delta<1$, we have

$$\|G'_{\bar{z}}(\ ,t)\|_p \le \|F^{-1}(\ ,t)\|_p + \|\mathrm{id}_{\Delta_\delta} F^{-2}(\ ,t)F'_{\bar{z}}(\ ,t)\|_p$$
$$\le A(\delta) + B(\delta,\tilde{p})\,\|F'_{\bar{z}}(\ ,t)\|_{p\tilde{p}},$$

where A and B are finite and depend on the indicated variables only, while $\tilde{p}$, $\tilde{p}>1$, can be chosen arbitrarily close to 1. It is worth-while to point out that the possibility of taking arbitrary $p>2$ is not needed now, but in order to assure the convergence of (13.1) in $B_p(\Delta)$ and thus p is fixed in this consideration.

In our case, the formula (4.5) in Lemma 4 is valid for $0<|z|<1$ and $0<t<T^*$; it reads

(13.10) $G(z,t) = \Phi^*(z,t) + T\,G'_{\bar{z}}(z,t),$

where Φ^* is holomorphic. Applying now the argument of Step G in the proof of Theorem 2 we obtain for $|z|<1$ and t, $t>0$, sufficiently small, an estimate analogous to (9.8), i.e.

(13.11) $|z-0| \le |\Psi^*(F(z,t),t) - \Psi^*(0,t)| + M|F(z,t)-0|^{1-2/p},\quad M<+\infty,$

where Ψ^* is holomorphic. It is clear that also Ψ^* satisfies a Hölder condition with the same exponent, so – in particular –

(13.12) $|\Psi^*(F(z,t),t) - \Psi^*(0,t)| \le M^*|F(z,t)-0|^{1-2/p},\quad M^*<+\infty.$

Relations (13.9), (13.11), and (13.12) yield

$$(13.13)\quad |G(z,t)| \le (M+M^*)^{1/(1-2/p)} |z|^{1-1/(1-2/p)} .$$

If t, $t>0$, is sufficiently small we can take here, by Theorem 1, $p>4$. In our consideration p, $p>2$, has to be arbitrary and fixed, so, to be more precise, we may say that for t, $t>0$, sufficiently small, we can replace p by some p^*, $p^*>4$, and – owing to Hölder's inequality – this gives no loss of generality when proving the convergence of (13.1) in $B_p(\Delta)$.

The inequality $p^*>4$ yields $1-1/(1-2/p^*)>-1$. Hence, by (13.1),

$$(13.14)\quad \int_{|z|=\delta} |G(z,t)|\,|1-\bar{s}z|^{-1}\,|dz| \longrightarrow 0 \quad \text{as} \quad \delta \longrightarrow 0+.$$

Since $T\,G_{\bar{z}}'(z,t)$ tends to a finite limit as $t \longrightarrow 0+$, then also

$$(13.15)\quad \int_{|z|=\delta} |T\,G_{\bar{z}}'(z,t)|\,|1-\bar{s}z|^{-1}\,|dz| \longrightarrow 0 \quad \text{as} \quad \delta \longrightarrow 0+.$$

Relations (13.10), (13.14), and (13.15) yield that the point $z=0$ is a removable singularity of Φ^* and, consequently,

$$(13.16)\quad \int_{\partial\Delta^r} \frac{G(z,t)}{1-\bar{s}z}\,dz = \int_{\partial\Delta^r} \frac{T\,G_{\bar{z}}'(z,t)}{1-\bar{s}z}\,dz, \quad |s|<r<1.$$

Next, similarly as in Step A, by Hölder's inequality, for any annulus Δ_δ, $0<\delta<1$, we get

$$|T\,G_{\bar{z}}'(z,t)| \le (1/\pi)\,[2\pi/(2-p')]^{1/p'}\, 2^{(2-p')/p}\, \|G_{\bar{z}}'(\ ,t)\|_p ,$$

where $1/p+1/p'=1$. On the other hand $T\,G_{\bar{z}}'(\ ,t)$ is holomorphic outside of ^{r} and vanishes at infinity, so, by the classical Cauchy integral formula, we obtain

$$\int_{\partial\Delta^r} \frac{G(z,t)}{1-\bar{s}z}\,dz = \begin{cases} 2\pi i\, s^{-1}\, T\,G_{\bar{s}}'(1/\bar{s},t) & \text{for } s\neq 0,\\ 0 & \text{for } s=0. \end{cases}$$

Therefore, by (13.7), (13.9), and (13.16), relation (13.8) follows, as desired.

S t e p D. __In turn we shall express the first two addends on the right-hand side of__ (13.6) __in terms of__ F __as double integrals using the normalization conditions__ $F(0,t)=0$ __and__ $F(1,t)=1$, $0<t<T^*$. First of all we observe that the formula (13.2), where Φ is defined by (13.6), Ψ by (13.7), and $T\,F_{\bar{s}}'$ by (13.3), remains valid for $|s|=1$.

From (13.2), (13.6), and (13.3) with $s=0$ we get

$$(13.17)\quad \int_{\partial\Delta} \frac{F(z,t)}{z}\,dz = 2i\iint_{\Delta} \frac{F_{\bar z}(z,t)}{z}\,dxdy.$$

Similarly, (13.2), (13.6), (13.8), and (13.3) with $s=1$ yield

$$(13.18)\quad \int_{\partial\Delta} \frac{F(z,t)}{z^2}\,dz = 2\pi i - 2i\iint_{\Delta} \frac{F_{\bar z}(z,t)}{z}\,dxdy$$

$$- 2i\iint_{\Delta} \frac{\partial}{\partial z}\frac{1/F(z,t)}{1-\bar z}\,\bar z\,dxdy + 2i\iint_{\Delta}\frac{F_{\bar z}(z,t)}{z-1}\,dxdy.$$

Step E. Finally we are going to prove the existence and to calculate the limit (13.1) in $B_p(\Delta)$. From (13.2), (13.6), (13.17), (13.18), and (13.3) we have

$$F(s,t) = s - \frac{1}{\pi}\iint_{\Delta} F_{\bar z}(z,t)\left(-\frac{1}{z} + \frac{s}{z} - \frac{s}{z-1} + \frac{1}{z-s}\right)dxdy$$

$$- \frac{1}{\pi}\iint_{\Delta} \bar F_z^{-1}(z,t)\left(\frac{s\bar z}{1-\bar z} - \frac{s^2\bar z}{1-s\bar z}\right)dxdy$$

$$= s + \frac{s(1-s)}{\pi}\iint_{\Delta}\left[\frac{F_{\bar z}(z,t)}{z(1-z)(s-z)} - \frac{\bar z\,\bar F_z^{-1}(z,t)}{(1-\bar z)(1-s\bar z)}\right]dxdy,$$

where $\bar F_z^{-1} = (\bar F^{-1})_z$. Hence

$$(13.19)\quad \frac{F(s,t)-s}{t} = \frac{s(1-s)}{\pi}\iint_{\Delta}\left[\frac{(1/t)F_{\bar z}(z,t)}{z(1-z)(s-z)} - \frac{(1/t)\bar z\,\bar F_z^{-1}(z,t)}{(1-\bar z)(1-s\bar z)}\right]dxdy.$$

By the hypotheses, $\mu(\ ,t) = t[\varphi + \varepsilon(\ ,t)]$, where $\|\varepsilon(\ ,t)\|_\infty \le \text{const}$ and $\varepsilon(\ ,t) \to 0$ as $t \to 0+$ a.e. in Δ. Thus, since $F'_{\bar s} = F'_s$, the formula (13.19) becomes

$$\frac{F(s,t)-s}{t} = \frac{s(1-s)}{\pi}\iint_{\Delta}\left[\frac{\varphi(z)+\varepsilon(z,t)}{z(1-z)(s-z)}F_{\bar z}(z,t)\right.$$

$$\left. - \frac{\overline{\varphi(z)} + \overline{\varepsilon(z,t)}}{(1-\bar z)(1-s\bar z)}\,\bar z\,\bar F_{\bar z}^{-1}(z,t)\right]dxdy$$

$$= \frac{s(1-s)}{\pi}\iint_{\Delta}\left[\frac{\varphi(z)}{z(1-z)(s-z)} + \frac{\overline{\varphi(z)}}{\bar z(1-\bar z)(1-s\bar z)}\right]dxdy$$

$$+\frac{s(1-s)}{\pi}\iint_\Delta\Big[\frac{\varepsilon(z,t)}{z(1-z)(s-z)}+\frac{\overline{\varepsilon(z,t)}}{\bar z(1-\bar z)(1-s\bar z)}\Big]dxdy$$

$$+\frac{s(1-s)}{\pi}\iint_\Delta\frac{\varphi(z)+\varepsilon(z,t)}{z(1-z)(s-z)}[F_z(z,t)-1]dxdy$$

$$+\frac{s(1-s)}{\pi}\iint_\Delta\frac{\overline{\varphi(z)}+\overline{\varepsilon(z,t)}}{\bar z(1-\bar z)(1-s\bar z)}\frac{\bar z^2}{\bar F^2(z,t)}[\bar F_{\bar z}(z,t)-1]dxdy$$

$$-\frac{s(1-s)}{\pi}\iint_\Delta\frac{\overline{\varphi(z)}+\overline{\varepsilon(z,t)}}{\bar z(1-\bar z)(1-s\bar z)}\frac{\bar F^2(z,t)-\bar z^2}{\bar F^2(z,t)}dxdy.$$

It remains to prove that the difference between $(1/t)[F(s,t)-s]$ and the expected limit (13.1) is arbitrarily small in $B_p(\Delta)$ when t is sufficiently small. This difference consists of four integrals, I_1, I_2, I_3, and I_4, say. The integral I_1 tends to 0 in $B_p(\Delta)$ as $t\to 0+$ since $\|\varepsilon(\ ,t)\|_\infty\le$ const and $\varepsilon(\ ,t)\to 0$ as $t\to 0+$ a.e. in Δ. By Lemma 15 and (1.2) the same concerns I_2 and I_3. Finally, taking into account I_4, we observe that, by (13.2), (13.3), and $F(s,0)=s$ for $|s|\le 1$, we have

(13.20) $$|F(s,t)-s|\le|\Phi(s,t)-\Phi(s,0)|+\frac{1}{\pi}\iint_\Delta|F_{\bar z}'(z,t)|\,|z-s|^{-1}dxdy$$

Continuing $F(\ ,t)$ to a mapping $F^*(\ ,t)\in S_Q^*$ by the formulae

$$F^*(s,t)=1/\overline{F(1/\bar s,\ t)}\quad\text{for}\quad 1<|s|<+\infty,\quad F^*(\infty,t)=\infty,$$

we can easily show, arguing as in the proof of Lemma 15, that it is sufficient to replace in our problem the Φ in question by $\Phi=\mathrm{id}_\Delta$. Since, by the hypothesis, $F_{\bar z}'(z,t)=t[\ (z)+\ (z,t)]F_z'(z,t)$, then finally (13.20) becomes

$$|F(s,t)-s|\le(1/\pi)t\iint_\Delta|\varphi(z)|\,|F_z'(z,t)-1|dxdy$$

$$+(1/\pi)t\iint_\Delta(|\varphi(z)|+|\varepsilon(z,t)|\,|F_z'(z,t)|)dxdy.$$

This means that also I_4 tends to 0 in $B_p(\Delta)$ as $t\to 0+$.

Thus we have completed the proof of (13.1) and the estimate (13.11) shows that the exponent $1-2/p$ in the norm in question cannot be improved with the same reasoning.

14. The parametrical equation for mappings of the unit disc

We now proceed to prove the first basic theorem on parametrization, announced in Section 12, which reads as follows:

THEOREM 5. Suppose that the functions $f(\ ,t)$, $0 \le t \le T$, belong to $S_{Q(t)}$ as functions generated by the complex dilatations $\mu(\ ,t)$ having the partial derivative μ_t a.e. in , where Q is given by (9.2). Then $w = f(s,t)$ is a solution of the differential equation

$$(14.1) \qquad w_t = \frac{w(1-w)}{\pi} \iint_\Delta \left[\frac{\varphi(z,t)}{z(1-z)(w-z)} + \frac{\overline{\varphi(z,t)}}{\bar{z}(1-\bar{z})(1-w\bar{z})} \right] dxdy,$$

where $x = \operatorname{re} z$, $y = \operatorname{im} z$, $\iint_D = \lim_{\varepsilon \to 0+} \iint_{D \setminus \Delta^\varepsilon(w)}$, and

$$(14.2) \qquad \varphi(w,t) = \{1/(1-|\mu(\check{f}(w,t),t)|^2)\}\, \mu_t(\check{f}(w,t),t) \times \exp\{-2i \arg \check{f}_w(w,t)\}.$$

Proof. For greater clarity the proof is divided into four steps.

Step A. In order to find a differential equation for functions belonging to the class $S_{Q(t)}$ and to apply for this Lemma 16, we construct a suitable function satisfying the assumptions of this lemma; we denote this function by F. By a suitable function we understand any function of the variable w, $|w| \le 1$, depending on two real parameters t, $0 \le t \le T$, and τ, $0 < \tau < \tau^*$, and fulfilling the condition

$$(14.3) \qquad (1/\tau)[F(w,t,\tau) - w] \to f_t(g(w,t),t) \quad \text{as} \quad \tau \to 0+$$

in any Banach space $B_p(\Delta)$, defined in Section 13, where g is chosen so that $F(\ ,\ ,\tau)$ belong to some $S_{Q^*(\tau)}$, $1 \le Q^*(\tau) < +\infty$.

The simplest way is to choose F so that the expression on the left-hand side of (14.3) be a difference quotient corresponding to f_t, i.e. so that

$$(14.4) \qquad F(w,t,\tau) = f(g(w,t), t+\tau),$$

$$(14.5) \qquad w = f(g(w,t),t).$$

Since $f(\ ,t)$ are invertible as belonging to $S_{Q(t)}$, then (14.5) yields $g = \check{f}$. By (14.4) this means that the most convenient is to admit

$$(14.6) \qquad F(w,t,\tau) = f(\check{f}(w,t), t+\tau).$$

It remains to verify whether the function F, defined by (14.6) for $|w| \leq 1$, $0 \leq t \leq T$, and $0 < \tau < T - t$, satisfies the assumptions of Lemma 16.

S t e p B. For editorial reasons let us begin with evaluation of the functions ε and φ, defined by

$$(14.7) \quad \varepsilon(w,t,\tau) + \varphi(w,t) = F'_{\bar{w}}(w,t,\tau)/\tau\, F'_{w}(w,t,\tau), \qquad |w| \leq 1,\ 0 \leq t \leq T,\ 0 < \tau < T - t,$$

$$(14.8) \quad \varepsilon(w,t,\tau) \to 0 \text{ as } \tau \to 0+ \text{ a.e. in } \Delta,\ 0 \leq t \leq T,$$

in dependence on f and μ. The distributional derivatives in (14.7) exist because of (14.6) and $f \in S_Q$.

By Lemmas 1, 3, and 14 these derivatives are given by the formulae

$$(14.9) \quad F'_{w}(w,t,\tau) = f'_z(\check{f}(w,t), t+\tau)\check{f}'_w(w,t) + f'_{\bar{z}}(w,t), t+\tau)\overline{\check{f}}'_w(w,t) \quad \text{a.e.},$$

$$(14.10) \quad F'_{\bar{w}}(w,t,\tau) = f'_z(\check{f}(w,t), t+\tau)\check{f}'_{\bar{w}}(w,t) + f'_{\bar{z}}(w,t), t+\tau)\overline{\check{f}}'_{\bar{w}}(w,t) \quad \text{a.e.}$$

By the same lemmas applied to the function $f \circ \check{f} = \mathrm{id}_\Delta$, we have

$$(f'_z \circ \check{f})\check{f}'_w + (f'_{\bar{z}} \circ \check{f})\overline{\check{f}}'_w = 1 \quad \text{and} \quad (f'_z \circ \check{f})\check{f}'_{\bar{w}} + (f'_{\bar{z}} \circ \check{f})\overline{\check{f}}'_{\bar{w}} = 0 \quad \text{a.e.},$$

whence, by (1.2) and (1.5),

$$\check{f}'_w = (\overline{f'_z} \circ f)/(J_f \circ f) \quad \text{and} \quad \overline{\check{f}}'_w = -(\overline{f'_{\bar{z}}} \circ f)/(J_f \circ f) \quad \text{a.e.},$$

$$\check{f}'_{\bar{w}} = -(f'_{\bar{z}} \circ f)/(J_f \circ f) \quad \text{and} \quad \overline{\check{f}}'_{\bar{w}} = (f'_z \circ f)/(J_f \circ f) \quad \text{a.e.}$$

Consequently, (14.7), (14.9), and (14.10) yield

$$\varepsilon(w,t,\tau) + \varphi(w,t)$$
$$= \frac{1}{\tau}\,\frac{-f'_z(\check{f}(w,t), t+\tau)f'_{\bar{z}}(\check{f}(w,t), t) + f'_{\bar{z}}(\check{f}(w,t), t+\tau)f'_z(\check{f}(w,t), t)}{f'_z(\check{f}(w,t), t+\tau)\overline{f'_{\bar{z}}}(\check{f}(w,t), t) - f'_{\bar{z}}(\check{f}(w,t), t+\tau)\overline{f'_{\bar{z}}}(\check{f}(w,t), t)} \quad \text{a.e.}$$

Hence, by the definition of $f(\ ,t)$ as a member of $S_{Q(t)}$ generated by the complex dilatation $\mu(\ ,t)$, we get

$$(14.11) \quad \varepsilon(w,t,\tau) + \varphi(w,t)$$
$$= \frac{1}{\tau}\,\frac{\mu(\check{f}(w,t), t+\tau) - \mu(\check{f}(w,t), t)}{1 - \mu(\check{f}(w,t), t+\tau)\overline{\mu(\check{f}(w,t), t)}}\exp\{2i \arg f'_z(\check{f}(w,t), t)\} \quad \text{a.e.}$$

Finally, by (14.8), we obtain

$$(14.12)\quad \varphi(w,t) = \{1/(1-|\mu(\check{f}(w,t),t)|^2)\}\mu_t(\check{f}(w,t),t) \times \exp\{2i\arg f_z'(\check{f}(w,t),t)\} \quad \text{a.e.}$$

Step C. Now it is natural to verify that the function φ, given by (14.2), is identical a.e. to that given by (14.2). To this end it is sufficient to check that

$$(14.13)\quad \operatorname{Arg} f_z'(\check{f}(w,t),t) = -\operatorname{Arg}\check{f}_w(w,t) \quad \text{a.e.}$$

This is easily verified by applying, as in Step B, Lemmas 1, 3, and 14 to the function $\check{f}\circ f = \mathrm{id}_\Delta$, whence

$$(14.14)\quad f_z'\circ\check{f} = \overline{\check{f}_w'}/J_{\check{f}} \quad \text{a.e.}$$

Since each $\check{f}(\ ,t)$ is $Q(t)$-qc, then $\operatorname{sgn} J_{\check{f}}(\ ,t) = 1$ at each regular point of Δ (cf. Section 1). Thus we obtain

$$\operatorname{Arg} f_z'\circ\check{f} = \operatorname{Arg}\overline{\check{f}_w'} \quad \text{a.e.}$$

On the other hand, by Corollary 5 and Lemma 3, $\check{f}_w' = \check{f}_w$ a.e. Therefore (14.3) follows, as desired.

In addition, owing to (14.12) and (14.13), we may express $\varepsilon(w,t,\tau)$, given by (14.11), in a more convenient form

$$(14.15)\quad \varepsilon(w,t,\tau) = \Big[\frac{1}{\tau}\,\frac{\mu(\check{f}(w,t),t+\tau)-\mu(\check{f}(w,t),t)}{1-\mu(\check{f}(w,t),t+\tau)\overline{\mu(\check{f}(w,t),t)}} - \frac{\mu_t(\check{f}(w,t),t)}{1-|\mu(\check{f}(w,t),t)|^2}\Big]\exp\{-2i\arg\check{f}_w'(w,t)\} \quad \text{a.e.}$$

Step D. Next we have to check that the functions F, ε, φ, defined by (14.6), (14.7), (14.8), and then expressed more conveniently by (14.6), (14.15), (14.2), satisfy the hypotheses of Lemma 16, but this is trivial. Finally, by (14.6), (14.7), and (14.8), relation (13.1) gives in our case (14.1), and this completes the proof.

COROLLARY 6. If in Theorem 5 we additionally assume that, for all $s\in\Delta$, $\mu(s,\ )$ is of the class C^1 or, even less, that

$$(14.16)\quad \|\mu_t(\ ,t+\tau)-\mu_t(\ ,t)\|_p \longrightarrow 0 \quad \text{as} \quad \tau\longrightarrow 0+ \quad \text{for} \quad p\geq 1,$$

then also $f(s,\ )$ is of the class C^1 for all $s\in\Delta$.

Proof. We follow Ahlfors [5], pp. 104-106. By (1.5), the change of variables $z \mapsto f(z,t)$ in (14.1) leads to the formula

$$(14.17)\quad w_t = \frac{w(1-w)}{\pi}\iint_{\Delta}\left[\frac{\phi(z,t)}{z^*(1-z^*)(w-z^*)} + \frac{\overline{\phi(z,t)}}{\bar{z}^*(1-\bar{z}^*)(1-w\bar{z}^*)}\right]dxdy,$$

where $w = f(s,t)$, $|s| \le 1$, and

$$(14.18)\quad \phi(s,t) = \mu_t(s,t)f_s^2(s,t), \quad z^* = f(z,t).$$

Hence, for any $\varepsilon > 0$ the difference $f_t(s,t+\tau) - f_t(s,t)$, $-t \le \tau \le T-t$, may be regarded as the sum of the following four integrals:

(a) $I_1(t) = -\dfrac{w(1-w)}{\pi}\displaystyle\iint_{\Delta^\varepsilon(s)\cap\Delta}\frac{\mu_t(z,t)f_s^2(z,t)}{z^*(1-z^*)(w-z^*)}dxdy$, where $w = f(s,t)$,

(b) $I_2(t) = -I_1(t+\tau)$,

(c) the analogue of $I_1(t) + I_2(t)$ obtained by replacing

$\mu_t f_z^2/z^*(1-z^*)(w-z^*)$ with $\overline{\mu_t f_z^2}/\bar{z}^*(1-\bar{z}^*)(1-w\bar{z}^*)$,

(d) the remaining terms.

The first integral tends to zero as $\varepsilon \to 0+$ uniformly on Δ since, by (1.5), the change of variables $z \mapsto \check{f}(z,t)$ leads to the estimate

$$|I_1(t)| \le \frac{|w(1-w)|}{\pi}\iint_{f[\Delta^\varepsilon(s)\cap\Delta,t)}\frac{|\mu_t(\check{f}(z,t),t)f_s^2(\check{f}(z,t),t)|}{|z(1-z)(w-z)|}dxdy,$$

where $f[E,t) = \{w\colon w = f(z,t),\ z \in E\}$; thus

$$|f[\Delta^\varepsilon(s)\cap\Delta,t)| = \iint_{\Delta^\varepsilon(s)\cap\Delta}(|f_z|^2 - |f_{\bar{z}}|^2)dxdy \to 0 \text{ as } \varepsilon \to 0+.$$

uniformly on Δ. The same concerns the second integral. The third integral trivially tends to zero as $\varepsilon \to 0+$. The fourth integral can be expressed as the sum of six integrals whose integrands in each case can be estimated by a constant multiplied by one of the following expressions:

$$|1/z^*(1-z^*)[f(s,t+\tau)-z^*] - 1/z^*(1-z^*)[f(s,t)-z^*]|,$$

$$|1/\bar{z}^*(1-\bar{z}^*)[1-f(s,t+\tau)\bar{z}^*] - 1/\bar{z}^*(1-\bar{z}^*)[1-f(s,t)\bar{z}^*]|,$$

$$|\mu_t(z,t+\ ) - \mu_t(z,t)|, \text{ and } |f_z^2(z,t+\tau) - f_z^2(z,t)|.$$

The fact that the first two expressions tend to zero as $\tau \to 0+$ is rather trivial. By the hypothesis, we also have (14.16). Finally, by Lemma 15 (together with Corollary 5 and Lemma 3), we have

$$\|f_z(\ ,t+\tau)-f_z(\ ,t)\|_p \leq \|f_z'(\ ,t+\tau)-1\|_p + \|f_z'(\ ,t)-1\|_p \to 0$$
$$\text{as } \tau \to 0+ \text{ for } p \geq 1$$

and this suffices to conclude the proof.

COROLLARY 7. *If we replace the normalization conditions*

(14.19) $f(0,t)=0,\ f(1,t)=1,\ 0<t\leq T,$

in Theorem 5 *and Corollary* 6 *by*

(14.20) $f(s_j,t)=s_j,\ s_j$ *distinct and* $|s_j|=1,\ j=1,2,3,\ 0<t\leq T,$

then the only change in the conclusions is to replace (14.1) *by*

$$(14.21)\quad w_t = \frac{1}{\pi}\iint_\Delta \Big[\frac{\varphi(z,t)}{w-z}\prod_{k=1}^{3}\frac{s_k-w}{s_k-z} + \frac{\overline{\varphi(z,t)}}{1-w\bar z}\prod_{k=1}^{3}\frac{s_k-w}{1-s_k\bar z}\Big]dxdy.$$

Proof. By Lemma 4, for every t, $0\leq t\leq T$, the differential equation (14.1) may be expressed in the form

$$(14.22)\quad w_t = \frac{1}{\pi}\iint_\Delta \varphi(z,t)\,(w-z)^{-1}\,dxdy + h(w,t) \quad \text{a.e. in } \Delta,$$

where $h(\ ,t)$ is holomorphic in $\operatorname{int}\Delta$. If we change the normalization conditions, we have to solve another boundary value problem for h. Arguing as in the preceding case we arrive, after quite technical calculations, at the required relation (14.21), where $w=f(s,t)$.

15. The converse problem

The converse parametrization problem has been studied by Shah Tao-shing [1] and Ławrynowicz [1, 5]. The result may be formulated as follows (Ławrynowicz [5]):

THEOREM 6. *Suppose that* φ *is a measurable function defined on* $\Delta\times[0;T]$ *and such that* $\|\varphi(\ ,t)\|\leq\frac{1}{2}$ *for* $0\leq t\leq T$. *Then there exists a unique solution* $w=f(s,t)$ *of* (14.1) *on* $\Delta\times[0;T]$, *subject to the initial condition* $f(s,0)=s$, *which for every* t *represents a mapping belonging to* $S_{Q(t)}$, *where* $Q(t)\leq\exp t$.

Proof. For greater clarity the proof is divided into seven steps.

S t e p A. We begin with proving the existence of a solution of (14.1) on $\Delta \times [0;T]$. Let $G(w,t)$ denote the right-hand side of (14.1). Consider the expression $|G(w_1,t) - G(w_2,t)|$ for $|w_1|, |w_2| \leq 1$. We have

$$(15.1)\qquad \frac{w(1-w)}{z(1-z)(w-z)} = \frac{1}{w-z} + \frac{w-1}{z} - \frac{w}{1-z},$$

$$(15.2)\qquad \frac{w(1-w)}{\bar{z}(1-\bar{z})(1-w\bar{z})} = \frac{-w^3}{1-w\bar{z}} + \frac{w(1-w)}{\bar{z}} + \frac{w}{1-\bar{z}}.$$

Since

$$\varphi(z,t)\Big[\frac{w_1-1}{z} + \frac{w_2-1}{z}\Big] + \overline{\varphi(z,t)}\Big[\frac{w_1(w_1-1)}{\bar{z}} - \frac{w_2(w_2-1)}{\bar{z}}\Big]$$

$$= \varphi(z,t)\frac{w_1-w_2}{z} + \overline{\varphi(z,t)}(w_1-w_2)\frac{1-(w_1+w_2)}{\bar{z}},$$

$$\overline{\varphi(z,t)}\Big[\frac{w_1^3}{1-w_1\bar{z}} - \frac{w_2^3}{1-w_2\bar{z}}\Big]$$

$$= \overline{\varphi(z,t)}\frac{(w_1-w_2)(w_1^2+w_1w_2+w_2^2-w_1^2w_2-w_1w_2^2)}{(1-w_1\bar{z})(1-w_2\bar{z})}.$$

Then, by the relations $\|\varphi(\ ,t)\|_\infty \leq \frac{1}{2}$ and $|w_1|, |w_2| \leq 1$, we obtain

$$(15.3)\qquad |G(w_1,t) - G(w_2,t)| \leq \frac{1}{2\pi}|w_1-w_2|\Big(\iint_\Delta \frac{dxdy}{|z-w_1||z-w_2|}$$

$$+ 4\iint_\Delta \frac{dxdy}{|z|} + 2\iint_\Delta \frac{dxdy}{|z-1|} + 5\iint_\Delta \frac{dxdy}{|1-w_1\bar{z}||1-w_2\bar{z}|}\Big).$$

Arguing as in the proof of Lemma 6, we get, for sufficiently small $|w_1 - w_2|$ the first integral at the right-hand side of (15.3) to be estimated by $4\pi[1+\frac{3}{2}\log(4/|w_1-w_2|)]$. Thus, by the continuity in w_1 and w_2, $w_1 \neq w_2$, it is estimated by $M_1 + 6\pi|\log|w_1-w_2||$ for $|w_1|, |w_2| \leq 1$, including $w_1 = w_2$, where M_1 is a finite constant. Next, it is easy to verify directly that the second and third integrals in (15.3) are bounded; let M_2 and M_3 denote these bounds, respectively. Finally, in order to estimate the fourth integral in (15.3), we distinguish two cases. If $|w_1|$ or $|w_2| \leq \frac{1}{2}$ and both are ≤ 1, then this integral is bounded by a constant M_4 as a function continuous in w_1

and w_2. If $|w_1|$ and $|w_2|$ are both $\geq \frac{1}{2}$ and ≤ 1, then we get an estimate $M_5 + 6 \cdot 2^2 \pi |\log|w_1 - w_2||$, analogous to that obtained in the case of the first integral. Hence we conclude that

$$(15.4) \quad |G(w_1,t) - G(w_2,t)| \leq M_6 |w_1 - w_2| (1 + M_7 |\log|w_1 - w_2||),$$

where $M_6 = (1/2\pi)[M_1 + 4M_2 + 2M_3 + 5\max(M_4,M_5)]$ and $M_7 = 126\pi/M_6$.

Now it is clear that any function $G(\ ,t)$ is <u>continuous</u> (in w and $\bar{w}$). Since it is also bounded, we can apply a well known theorem of Peano (cf. e.g. Petrovski [1], p. 37) in a generalized form due to Carathéodory [2], pp. 665-688 (cf. Coddington and Levinson [1], p. 43), where – with respect to t – the function G needs only to be measurable and bounded by a Lebesgue integrable function. Thus there exists at least one solution $w = f(s,t)$ of (14.1) on $\Delta \times [0;T]$, subject to the initial condition $f(s,0) = s$.

S t e p <u>B</u>. In order to prove <u>the uniqueness of the solution in question</u> we are going to verify that the hypotheses of a well known theorem of Osgood (cf. e.g. Petrovski [1], p. 42) are fulfilled. To this end let us denote by $H(|w_1 - w_2|)$ the right-hand side of (15.4). In order to apply the theorem of Osgood we verify easily that H is positive-valued and continuous for $|w_1 - w_2| > 0$, and that

$$\int_c^d [1/H(r)]dr \longrightarrow +\infty \quad \text{as} \quad c \longrightarrow 0+, \quad c < d.$$

Consequently, there exists at most one solution $w = f(s,t)$ of (14.1) on $\Delta \times [0;T]$, subject to the initial condition $f(s,0) = 0$, and – by Step A – this solution is unique.

S t e p <u>C</u>. By a known theorem on the dependence of solutions of differential equations on parameters (cf. Coddington and Levinson [1], p. 58), from the continuity (in w and $\bar{w}$) and boundedness of G (Step A) and the uniqueness of the solution in question (Step B) it follows that, for any fixed t, <u>the function</u> $f(\ ,t)$ <u>is continuous on</u> Δ.

S t e p <u>D</u>. We prove next that, for any fixed t, $f(\ ,t)$ <u>is a sense-preserving homeomorphism</u>. We apply again the theorem quoted after Goursat [1] in the proof of Lemma 13. By this theorem the number of solutions of the equation $f(s,t) = w_0(t)$, where w_0 is continuous, $|w_0(t)| \leq 1$, $0 \leq t \leq T$, is equal to

$$\frac{1}{2\pi} \int_{\partial\Delta} d\{\arg[f(s,t) - w_0(t)]\} = \frac{1}{2\pi i} \int_{\partial f[\Delta]} [w - w_0(t)]^{-1} dw,$$

i.e. to the index of the point $w_0(t)$ with respect to $f[\Delta]$, which will be denoted by $n(t, w_0(t))$. The function n is continuous in w_0, t being fixed, and $f(1,t)=1$ for $0 \le t \le T$; the latter relation being a consequence of

$$(15.5)' \quad f(s,t) = s + \int_0^t G(f(s,t), t)dt$$

and $G(1,t)=0$. Therefore $n(t, w_0(t)) = n(t,1)$ for $w_0(t)$ taken from Δ. But n is also continuous in t, so $n(t,1) = n(0,1) = 1$ and, consequently, $n(t, w_0(t)) = 1$. Summing up, for any t, $0 \le t \le T$, $f(\ ,t)$ has to be one-to-one and sense-preserving and, since it is continuous, it must be a sense-preserving homeomorphism.

S t e p E. In turn we are going to prove that any $f(\ ,t)$ transforms Δ onto itself. Since f is a homeomorphism, it is sufficient to verify that $|f(s,t)| = 1$ for $|s| = 1$ and $0 \le t \le T$.

To this end let us note that $\mathrm{re}[(1/w)G(w,t)] = 0$ for $|s| = 1$, where $w = f(z,t)$. Then, introducing the notation $\varepsilon = \bar{f} - 1/f$ and applying an easily verified identity

$$\mathrm{re}[(1/f)(\partial/\partial t)f] = \tfrac{1}{2}(\partial/\partial t)(f\varepsilon)/(1 + f\varepsilon),$$

we obtain, by (14.1) with $w = f(s,t)$, after letting $|s| \to 1-$,

$$(\partial/\partial t)\log[1 + f(s,t)\varepsilon(s,t)] = 0 \quad \text{for } |s| = 1,\ 0 \le t \le T.$$

From the above it follows that $1 + f(s,t)\varepsilon(s,t) = e^{c(s)}$, where c does not depend on t. Hence, by the definition of ε,

$$|f(s,t)|^2 = e^{c(s)} \quad \text{for } |s| = 1,\ 0 \le t \le T.$$

Now, taking into account the initial condition $f(s,0) = s$, we see that $c(s) = 0$ identically, and thus $|f(s,t)| = 1$ for $|s| = 1$.

S t e p F. We wish to prove now a stronger version of Step C: we claim that, for any fixed t, the inverse function $\check{f}(\ ,t)$ is ACL. To this end we have to verify that, for any rectangle U, $\mathrm{cl}\, U \subset \mathrm{int}\, \Delta$, with sides parallel to the coordinate axes it is absolutely continuous on almost all line segments in U which are parallel to either side of U.

In fact, take U and v so that $U \cap \{w: \mathrm{im}\, w = v\} \ne \emptyset$ and let $(u_k; \tilde{u}_k)$, $k = 1, \ldots, m$, be contained in $\{u: u + iv\ \ U\}$ and be disjoint. Since, as proved in Step D, $f(\ ,t)$ is one-to-one, and, by hypothesis, $f(s,0) = s$, the relation (15.5) may be written in the form

$$\check{f}(w,t) = w - \int_0^t G(w,t)\,dt.$$

Applying to $G(w,t)$ the estimate (15.4), we get

$$\sum_{k=1}^{m} |\check{f}(\tilde{u}_k + iv,\, t) - \check{f}(u_k + iv,\, t)|$$

$$\leq M_6 \sum_{k=1}^{m} (\tilde{u}_k - u_k)[1 + M_7 \log(\tilde{u}_k - u_k)] \leq M_6(1 + M_7)[\sum_{k=1}^{m} (\tilde{u}_k - u_k)]^2,$$

provided that $0 \leq \tilde{u}_k - u_k \leq \frac{1}{4}$. This means that $\check{f}$ is ACL, as desired, since in the above consideration the roles of u and v can be interchanged.

S t e p G. Finally we are going to complete the proof, i.e. to show that for every t, $0 \leq t \leq T$, the mapping $f(\ ,t)$ belongs to $S_{Q(t)}$, where $Q(t) \leq \exp t$.

Let us notice that, as remarked in Section 2, the function $\check{f}(\ ,t)$, being ACL, possesses a.e. the derivatives $\check{f}_w(\ ,t)$ and $\check{f}_{\bar{w}}(\ ,t)$. Therefore we may consider (14.2) as a differential equation with respect to $\sigma = \nu(w,t)$, where $\nu(w,t) = \mu(\check{f}(w,t), t)$:

(15.6) $\quad \sigma_t = \varphi(w,t)(1 - |\sigma|^2) \exp\{2i \arg \check{f}_w(w,t)\}$ a.e.

Let $G^*(\sigma,w,t)$ denote the right-hand side of (15.6). Since any function $G^*(\ ,w,t)$, considered on $\operatorname{int}\Delta$, is continuous and bounded, applying again the theorem of Carathéodory, mentioned in Step A, we conclude that there exists at least one solution $\sigma = \nu(w,t)$ of (15.6) on $\Delta \times [0; T]$ subject to the initial condition $\nu(w,0) = 0$.

Let

(15.7) $\quad \mu(s,t) = \nu(f(s,t), t), \quad |s| \leq 1, \quad 0 \leq t \leq T,$

(15.8) $\quad p = (1 + |\mu|)/(1 - |\mu|).$

Then

(15.9) $$\varphi(w,t) = \lim_{\tau \to 0+} \frac{1}{\tau} \cdot \frac{\mu(s,\, t+\tau) - \mu(s,t)}{1 - \mu(s,\, t+\tau)\overline{\mu(s,t)}} \exp(-2i \arg s_w)$$

$$= -\tfrac{1}{2}[p_t(s,t)/p(s,t)] \exp\{-2i[\arg s_w + \theta(s,t)]\},$$

where $\theta = \frac{1}{2} \arg \mu$ and $s = \check{f}(w,t)$. On the other hand, the initial condition $\nu(w,0) = 0$ yields $\mu(s,0) = 0$, whence $p(s,0) = 1$. Therefore, by the relation $\|\varphi(\ ,t)\|_\infty \leq \frac{1}{2}$, we obtain

II. The parametrical methods

(15.10) $p(s,t) = \exp\{2 \int_0^t \dfrac{p_\tau(s,\tau)}{p(s,\tau)} \, d\tau\} = \exp\{2 \int_0^t |\varphi(f(s,\tau), \tau)| \, d\tau\} \le \exp t$

except perhaps for w such that $|\varphi(w,t)| > \|\varphi(\ ,t)\|_\infty$ for some t. Thus, by Theorem 4, formulae (9 2) and (15.8), and the uniqueness of $f(\ ,t)$ for every fixed t, $0 \le t \le T$, the function belonging to $S_{\exp t}$ as the function generated by the complex dilatation $(\ ,t)$ must be identical with $f(\ ,t)$. This proves that $f(\ ,t)$ belongs to $S_{Q(t)}$, where $Q(t) \le \exp t$, as desired.

It is natural to ask under which conditions φ defined by (14.2) satisfies $\|\varphi(\ ,t)\|_\infty \le \frac{1}{2}$ and whether the estimate $Q(t) \le \exp t$ is precise. For homotopies determined by the equations

(15.11) $\mu(s,t) = - \dfrac{[p(s)]^{t/T} - 1}{[p(s)]^{t/T} + 1} \exp 2i\theta(s), \quad 1 \le p(s) \le Q = \exp T, \quad \theta$ real,

the first question was answered in Shah Tao-shing [1] (Theorem 3). We are going to answer both questions entirely.

THEOREM 7. Under the hypotheses of Theorem 5, if the partial derivative p_t of the dilatation (15.8) is positive a.e. in Δ and such that $\|p_t(\ ,t)\|_\infty \le M < +\infty$ for $0 \le t \le T$, then the change of variable t $= T_0\tau$, where $T_0 = 1/M$, leads to an analogue of (14.1), where $\varphi(z,t)$ is replaced by $T_0\varphi(z, T_0\tau)$ and $\|T_0\varphi(\ , T_0\tau)\|_\infty \le \frac{1}{2}$ for $0 \le \tau \le T/T_0$.

Proof. Since $T_0 = 1/M$, then the relation (15.9), where $\theta = \frac{1}{2} \times$ arg μ and $s = \check{f}(w,t)$, yields

$T_0|\varphi(w, T_0\tau)| = \frac{1}{2}(1/M) |p_t(s, T_0\tau)| / |p(s, T_0\tau)|.$

Therefore, utilizing the estimates $p(s,t) \ge 1$, $|s| \le 1$, $0 \le t \le T$, and $\|p_t(\ ,t)\|_\infty \le M < +\infty$, $0 \le t \le T$, we obtain $T_0|\varphi(w, T_0\tau)| \le \frac{1}{2}$, $0 \le \tau \le T/T_0$, for $|w| \le 1$ except perhaps for w such that $|\varphi(w, T_0\tau)| > \|\varphi(\ , T_0\tau)\|_\infty$ for some τ, as desired.

THEOREM 8. Under the hypotheses of Theorem 6 we have

(15.12) $Q(t) \ge \exp\{2 \int_0^t |\varphi(w,\tau)| \, d\tau\}$, where $w = f(s,\tau)$,

and

(15.13) $Q(t) \ge \exp\{2 \int_0^t |\varphi(f(\check{f}(w,t), \tau), \tau)| \, d\tau\}$

for $|w| \le 1$ except perhaps for w such that $|\varphi(w,t)| > \|\varphi(\ ,t)\|_\infty$ for some t.

Proof. Relation (15.12) is an immediate consequence of (9.2), (15.8), and (15.10). In order to arrive at (15.13) one has to replace in this consideration p by the corresponding dilatation $\tilde{p}$ of the inverse mapping: $\tilde{p}(w,t) = p(\check{f}(w,t), t)$ and to observe that $\|p(\ ,t)\|_\infty = Q(t)$, $0 \le t \le T$.

Finally, in analogy to Corollary 7, after quite technical calculations, we obtain

COROLLARY 8. If we replace the normalization conditions (14.19) in Theorems 6-8 and Corollary 7 by (14.20), then the only additional change in the conclusions is to replace (14.1) by (14.21).

16. Parametrization in an annulus

We are going to prove the analogues of Theorems 4-7 for an annulus (Ławrynowicz [5]).

THEOREM 9. Suppose that the functions $f(\ ,t)$, $0 \le t \le T$, belong to $S_{Q(t)}^{r,R(t)}$ as functions generated by the complex dilatations $\mu(\ ,t)$ having the partial derivative μ_t a.e. in Δ_r, where Q is given by (9.2). Then $w = f(s,t)$ is a solution of the differential equation

$$(16.1)\quad w_t = \frac{w}{2\pi}\iint_{\Delta_{R(t)}} \sum_{n=-\infty}^{+\infty}\Big\{ \frac{\varphi(z,t)}{z^2}\Big[\frac{w + R^{2n}(t)z}{w - R^{2n}(t)z} - \frac{1 + R^{2n}(t)z}{1 - R^{2n}(t)z}\Big] - \frac{\overline{\varphi(z,t)}}{\bar{z}^2}\Big[\frac{1 + R^{2n}(t)w\bar{z}}{1 - R^{2n}(t)w\bar{z}} - \frac{1 + R^{2n}(t)\bar{z}}{1 - R^{2n}(t)\bar{z}}\Big]\Big\}dxdy,$$

where $x = \operatorname{re} z$, $y = \operatorname{im} z$, $R^{2n}(t) = [R(t)]^{2n}$, the notation $\ldots + a_{-1} + a_0 + a_1 + \ldots$ is applied for the sake of simplicity instead of $a_0 + (a_1 + a_{-1}) + \ldots$ provided that the last series converges, and φ is defined by (14.2). Moreover, $\rho = R(t)$, $0 \le t \le T$, is a differentiable function which satisfies the differential equation

$$(16.2)\quad \rho' = (1/2\pi)\iint_{\Delta_\rho} \rho[\varphi(z,t)/z^2 + \overline{\varphi(z,t)}/\bar{z}^2]dxdy.$$

Proof. For greater clarity the proof is divided into four steps.

Step A. We begin with continuing $Q(t)$-quasiconformally the mappings $f(\ ,t)$, $0 \le t \le T$, into the inner disc Δ_r in order to apply Theorem 5:

(16.3) $f^*(s,t) = R^{2n}(t)/\overline{f(r^{2n}/\bar{s},\, t)}$ for $r^{2n} \leq |s| < r^{2n-1}$, $n = 1,2,\ldots,$

(16.4) $f^*(s,t) = R^{2n}(t)\, f(s/r^{2n},\, t)$ for $r^{2n+1} \leq |s| < r^{2n}$, $n = 1,2,\ldots$

Obviously, we admit $f^*(s,t) = f(s,t)$ for $r \leq |s| \leq 1$ and $f^*(0,t) = 0$. It is easy to see that for any t, $0 \leq t \leq T$, the function $f^*(\ ,t)$ belongs to $S_Q(t)$ as the function generated by the complex dilatation $\mu^*(\ ,t)$ which is determined by the formulae

(16.5) $\mu^*(s,t) = e^{4i \arg s}\mu(r^{2n}/\bar{s},\, t)$ for $r^{2n} \leq |s| < r^{2n-1}$, $n = 1,2,\ldots,$

(16.6) $\mu^*(s,t) = \mu(s/r^{2n},\, t)$ for $r^{2n+1} < |s| < r^{2n}$, $n = 1,2,\ldots$

and, obviously, $\mu^*(s,t) = \mu(s,t)$ for $r \leq |s| \leq 1$. Hence, by Theorem 5, $w = f^*(s,t)$ is a solution of the differential equation (14.1), where φ is replaced with φ^* given by

$$(16.7)\quad \varphi^*(w,t) = \{1/(1 - |\mu^*(\check{f}^*(w,t),\, t)|^2\}\mu^*_t(\check{f}^*(w,t),\, t) \times \exp\{-2i \arg \check{f}^*_w(w,t)\}.$$

Step $\underline{B}$. We proceed to verify that $w = f(s,t)$ <u>is a solution of</u>

$$(16.8)\quad w_t = \frac{w}{\pi}\iint_{\Delta_{R(t)}} \Big[\frac{\varphi(z,t)}{z^2}\Big(-\frac{1}{1-w/z} + \frac{1}{1-1/z}\Big) - \frac{\overline{\varphi(z,t)}}{\bar{z}^2}\Big(\frac{1}{1-wz} - \frac{1}{1-\bar{z}}\Big)\Big]dxdy + \frac{w}{\pi}\sum_{n=1}^{+\infty}\iint_{\Delta_{R(t)}}\Big[\frac{\varphi(z,t)}{z^2}\Big(-\frac{1}{1-w/R^{2n}(t)z} + \frac{1}{1-1/R^{2n}(t)z} - \frac{1}{1-R^{2n}(t)w/z} + \frac{1}{1-R^{2n}(t)/z}\Big) - \frac{\overline{\varphi(z,t)}}{\bar{z}^2}\Big(\frac{1}{1-R^{2n}(t)w\bar{z}} - \frac{1}{1-R^{2n}(t)\bar{z}} + \frac{1}{1-w\bar{z}/R^{2n}(t)} - \frac{1}{1-\bar{z}/R^{2n}(t)}\Big)\Big]dxdy,$$

<u>where</u> φ <u>is defined by</u> (14.2).

To this end, starting with the formula (16.7), we get, by (16.3) -(16.6),

(16.9) $\varphi^*(w,t) = e^{4i\arg w}\,\overline{\varphi(R^{2n}(t)/\bar{w},\, t)}$ for $R^{2n}(t) \leq |w| < R^{2n-1}(t)$, $n = 1,2,\ldots,$

(16.10) $\varphi^*(w,t) = \varphi(w/R^{2n}(t),\, t)$ for $R^{2n+1}(t) \leq |w| < R^{2n}(t)$, $n = 1,2,\ldots$

Therefore, taking into consideration the relation (14.1), where $w = f^*(s,t)$ and φ is replaced with φ^* defined by (16.7), we can rearrange it as follows:

$$w_t = \frac{w(1-w)}{\pi} \iint_{\Delta_{R(t)}} \left[\frac{\varphi^*(z,t)}{z(1-z)(w-z)} + \frac{\overline{\varphi^*(z,t)}}{\bar{z}(1-\bar{z})(1-w\bar{z})} \right] dxdy$$

$$+ \frac{w(1-w)}{\pi} \sum_{n=1}^{+\infty} \Big\{ \iint_{\Delta(2n)} + \iint_{\Delta(2n-1)} \Big\} \left[\frac{\varphi^*(z,t)}{z(1-z)(w-z)} + \frac{\overline{\varphi^*(z,t)}}{\bar{z}(1-\bar{z})(1-w\bar{z})} \right] dxdy$$

$$= \frac{w(1-w)}{\pi} \iint_{\Delta_{R(t)}} \left[\frac{\varphi(z,t)}{z(1-z)(w-z)} + \frac{\overline{\varphi(z,t)}}{\bar{z}(1-\bar{z})(1-w\bar{z})} \right] dxdy$$

$$+ \frac{w(1-w)}{\pi} \sum_{n=1}^{+\infty} \Big\{ \iint_{\Delta(2n)} \left[\frac{\varphi(z/R^{2n}(t), t)}{z(1-z)(w-z)} + \frac{\overline{\varphi(R^{2n}(t)/\bar{z}, t)}}{\bar{z}(1-\bar{z})(1-w\bar{z})} \right] dxdy$$

$$+ \iint_{\Delta(2n-1)} \left[e^{4i \arg z} \frac{\overline{\varphi(R^{2n}(t)/\bar{z}, t)}}{z(1-z)(w-z)} + e^{-4i \arg z} \frac{\varphi(R^{2n}(t)/\bar{z}, t)}{\bar{z}(1-\bar{z})(1-w\bar{z})} \right] dxdy \Big\},$$

where $w = f(s,t)$, $r \le |s| \le 1$, and

(16.11) $\Delta(2n-1) = \{w:\ R^{2n}(t) \le |w| \le R^{2n-1}(t)\}$, $\Delta(2n) = \{w:\ R^{2n+1}(t) \le |w| \le R^{2n}(t)\}$.

Now, the change of variables $z \mapsto R^{2n}(t)/\bar{z}$ in the integral over $\Delta(2n-1)$ and $z \mapsto z/R^{2n}(t)$ in the integral over $\Delta(2n)$, $n = 1,2,\ldots$, leads to the following relation:

$$w_t = \frac{w(1-w)}{\pi} \iint_{\Delta_{R(t)}} \left[\frac{\varphi(z,t)}{z(1-z)(w-z)} + \frac{\overline{\varphi(z,t)}}{\bar{z}(1-\bar{z})(1-w\bar{z})} \right] dxdy$$

$$+ \frac{w(1-w)}{\pi} \sum_{n=1}^{+\infty} \iint_{\Delta_{R(t)}} R^{2n}(t) \Big\{ \frac{\varphi(z,t)}{z[1-R^{2n}(t)z][w-R^{2n}(t)z]}$$

$$+ \frac{\overline{\varphi(z,t)}}{\bar{z}[1-R^{2n}(t)\bar{z}][1-R^{2n}(t)w\bar{z}]} + \frac{\overline{\varphi(z,t)}}{\bar{z}[\bar{z}-R^{2n}(t)][w\bar{z}-R^{2n}(t)]}$$

$$+ \frac{\varphi(z,t)}{z[z-R^{2n}(t)][z-R^{2n}(t)w]} \Big\} dxdy,$$

i.e. (16.8), where $w=f(s,t)$, $r\leq|s|\leq 1$, and φ is defined by (14.2).

Step C. In turn we are going to rearrange the relation (16.8), where $w=f(s,t)$, $r\leq|s|\leq 1$, and φ is defined by (14.2), to the required form (16.1).

It can be easily checked with help of the well known Weierstrass' test that the series of integrands in (16.8) is uniformly convergent. Indeed, let us notice that

$$\frac{1}{1-R^{2n}w/z}=1-\frac{1}{1-z/R^{2n}w},\quad \frac{1}{1-R^{2n}w\bar{z}}=1-\frac{1}{1-1/R^{2n}w\bar{z}}.$$

Since $R(t)\leq|z|\leq 1$, $R(t)\leq|w|\leq 1$ and $0<R(t)<1$ for $0\leq T\leq 1$, we have

$$|1-z/R^{2n}w|\geq|z/R^{2n}w|-1\geq R^{1-2n}-1>R^{1-2n}(1-R),$$

$$|1-w/R^{2n}z|\geq|w/R^{2n}z|-1\geq R^{1-2n}-1>R^{1-2n}(1-R),$$

$$|1-w\bar{z}/R^{2n}|\geq|w\bar{z}/R^{2n}|-1\geq R^{2-2n}-1>R^{2-2n}(1-R),$$

$$|1-1/R^{2n}w\bar{z}|\geq|1/R^{2n}w\bar{z}|-1\geq R^{-2n}-1>R^{-2n}(1-R).$$

Consequently

$$\left|\frac{1}{1-w/R^{2n}z}-\frac{1}{1-R^{2n}w/z}+1\right|<\frac{2R^{2n-1}}{1-R},$$

$$\left|\frac{1}{1-w\bar{z}/R^{2n}}-\frac{1}{1-R^{2n}w\bar{z}}+1\right|<\frac{2R^{2n-2}}{1-R}.$$

Hence the n-th integrand in question is estimated by

$$4R^{2n-4}(t)\max_{z\in\Delta_{R(t)}}|\varphi(z,t)|$$

which suffices to prove the desired uniform convergence.

Since all integrals appearing in (16.8) are continuous in w and $\bar{w}$ (cf. the proof of Theorem 6, Step A), then for any w, $r\leq|w|\leq 1$, we can interchange the order of integration and summation in (16.8). In order to simplify further (16.8) we subtract $\frac{1}{2}$ from each fraction appearing in the parantheses (the number of fractions with the signs + and - is always the same). In this way we obtain the relation

$$w_t=\frac{w}{2\pi}\iint_{\Delta_{R(t)}}\left[\frac{\varphi(z,t)}{z^2}\left(-\frac{1+w/z}{1-w/z}+\frac{1+1/z}{1-1/z}\right)-\frac{\overline{\varphi(z,t)}}{\bar{z}^2}\left(\frac{1+w\bar{z}}{1-w\bar{z}}-\frac{1+\bar{z}}{1-\bar{z}}\right)\right]$$

$$\times dxdy + \frac{w}{2\pi}\iint_{\Delta_{R(t)}} \sum_{n=1}^{+\infty}\{\frac{\varphi(z,t)}{z^2}[-\frac{1+w/R^{2n}(t)z}{1-w/R^{2n}(t)z}+\frac{1+1/R^{2n}(t)z}{1-1/R^{2n}(t)z}$$

$$-\frac{1+R^{2n}(t)w/z}{1-R^{2n}(t)w/z}+\frac{1+R^{2n}(t)/z}{1-R^{2n}(t)/z}]-\frac{\overline{\varphi(z,t)}}{\bar{z}^2}[\frac{1+R^{2n}(t)w\bar{z}}{1-R^{2n}(t)w\bar{z}}$$

$$-\frac{1+R^{2n}(t)z}{1-R^{2n}(t)\bar{z}}+\frac{1+w\bar{z}/R^{2n}(t)}{1-w\bar{z}/R^{2n}(t)}-\frac{1+\bar{z}/R^{2n}(t)}{1-\bar{z}/R^{2n}(t)}]\}dxdy$$

i.e. (16.1), where $w=f(s,t)$, $r\le|s|\le 1$.

S t e p D. Finally we verify that $\rho=R(t)$, $0\le t\le T$, <u>is a differentiable function which satisfies the differential equation</u> (16.2). To this end let us note that, by (16.1),

$$\lim_{\tau\to 0+}\frac{1}{\tau}[R(t+\tau)-R(t)]=R(t)\,\mathrm{re}[\frac{1}{w}G(w,t)] \quad \text{for } |s|=r,$$

where $G(w,t)$ denotes the right-hand side of (16.1) and $w=f(s,t)$. Since, as it can easily be verified, for $|w|=R(t)$ we have

$$\mathrm{re}[(1/w)G(w,t)]=(1/2\pi)\iint_{\Delta_{R(t)}}[\varphi(z,t)/z^2+\overline{\varphi(z,t)}/\bar{z}^2]dxdy,$$

then the desired conclusion follows and this completes the proof.

<u>COROLLARY 9. If in Theorem</u> 9 <u>we additionally assume that, for all</u> $s\in\Delta_r$, $\mu(s,\ )$ <u>is of the class</u> C^1 <u>or, even less, that</u> (14.16) <u>holds, the integral being taken over</u> Δ_r, <u>then also</u> $f(s,\ )$ <u>and</u> R <u>are of the class</u> C^1 <u>for all</u> $s\in\Delta_r$.

<u>P r o o f</u>. The first method is to proceed exactly analogously to the proof of Corollary 6. Then again the crucial fact is that the change of variables $z\mapsto f(z,t)$ in (16.1) leads to the formula

$$(16.12)\quad w_t=\frac{w}{2\pi}\iint_{\Delta_r}\sum_{n=-\infty}^{+\infty}\{\frac{\phi(z,t)}{z^{*2}}[\frac{w+R^{2n}(t)z^*}{w-R^{2n}(t)z^*}-\frac{1+R^{2n}(t)z^*}{1-R^{2n}(t)\bar{z}^*}]$$

$$-\frac{\overline{\phi(z,t)}}{\bar{z}^{*2}}[\frac{1+R^{2n}(t)w\bar{z}^*}{1-R^{2n}(t)w\bar{z}^*}-\frac{1+R^{2n}(t)\bar{z}^*}{1-R^{2n}(t)\bar{z}^*}]\}dxdy,$$

where $w=f(s,t)$, $r\le|s|\le 1$, and ϕ and z^* are defined by (14.18). Since $R=|f(r,\ )|$, then also R is of the class C^1.

The second method is considerably simpler. We continue $Q(t)$-

quasiconformally the mappings $f(\ ,t)$, $0 \le t \le T$, into the inner disc Δ_r (cf. Step A in the proof of Theorem 8), easily prove that the continued complex dilatation μ^* fulfills the condition

$$\|\mu_t^*(\ ,t+\tau)-\mu_t^*(\ ,t)\|_p \longrightarrow 0 \text{ as } \tau \longrightarrow 0+ \text{ for } p \ge 1,$$

the integral being taken over Δ, and then apply Corollary 6 directly. Since $R = |f(r,\ )|$, then also R is of the class C^1.

THEOREM 10. *Suppose that* φ *is a measurable function defined on* $\Delta \times [0;T]$ *and such that* $\|\varphi(\ ,t)\|_\infty \le \frac{1}{2}$ *for* $0 \le t \le T$. *Then there exists a unique solution* $\rho = R(t)$ *of* (16.2) *on* $[0;T]$, *subject to the initial condition* $R(0) = r$. *Furthermore, there exists also a unique solution* $w = f(s,t)$ *of* (16.1) *on* $\Delta_r \times [0;T]$, *subject to the initial condition* $f(s,0) = s$, *which for every* t *represents a mapping belonging to* $S_{Q(t)}^{r,R(t)}$, *where* $Q(t) \le \exp t$.

Proof. For greater clarity the proof is divided into three steps.

Step A. We begin with proving *the existence of a unique solution* $\rho = R(t)$ *of* (16.2) *on* $[0;T]$, *subject to the initial condition* $R(0) = r$.

Let $H(\rho,t)$ denote the right-hand side of (16.2). Consider the expression $|H(\rho_1,t) - H(\rho_2,t)|$ for $R(t) \le \rho_1 \le 1$, $R(t) \le \rho_2 \le 1$. Since $\|\varphi(\ ,t)\|_\infty \le \frac{1}{2}$, it is bounded by a constant multiplied by $|\rho_1 - \rho_2|$. Now it is clear that any function $H(\ ,t)$ is continuous. Since it is also bounded, we can apply the already quoted (cf. Step A in the proof of Theorem 6) theorem of Peano in a generalized form due to Carathéodory, by which there exists at least one solution $\rho = R(t)$ of (16.2) on $[0;T]$, subject to the initial condition $R(0) = r$.

In order to prove the uniqueness of the solution in question we verify that the hypotheses of the already quoted theorem of Osgood are fulfilled what is quite easy because of the above estimate of $|H(\rho_1, t) - H(\rho_2,t)|$.

Step B. We are going to prove now *the existence of a unique solution* $w = f(s,t)$ *of* (16.1) *on* $\Delta_r \times [0;T]$, *subject to the initial condition* $f(s,0) = s$, *which for every* t *represents a* $Q(t)$-*qc mapping of the annulus* Δ_r *into* Δ.

We continue the function φ into the inner disc Λ_r by the formulae (16.9), (16.10). Then, by Theorem 6, there exists a unique solution $w = f^*(s,t)$ of (14.1), where φ is replaced with φ^*, defined on $\Delta \times [0;T]$ and subject to the initial condition $f^*(s,0) = s$, which for every t represents a mapping belonging to $S_{Q(t)}$, where $Q(t) \leq \exp t$. Now we set $f = f^*|\Delta_r$ and rearrange, as in the proof of Theorem 6 (Steps B and C) the relation (14.1), where $w = f^*(s,t)$, $|s| \leq 1$, and φ is replaced by φ^*, to the required form (16.1), where $w = f(s,t)$, $r \leq |s| \leq 1$.

S t e p C. It remains to prove that any $f(\ ,t)$ transforms Δ_r onto $\Delta_{R(t)}$ and $f(1,t) = 1$.

Let $G(w,t)$ denote the right-hand side of (16.1). The relation $|f(s,t)| = 1$ for $|s| = 1$ can be obtained exactly as in the proof of Theorem 6 (Step E) with the corresponding change of the meaning of G and with replacing (14.1) by (16.1).

In order to prove the relation $|f(s,t)| = R(t)$ for $|s| = r$ let us note that $\mathrm{re}[(1/w)G(w,t)] = R'(t)/R(t)$ for $|s| = r$, where $w = f(z,t)$. Then, introducing the notation $\varepsilon = \bar{f} - R^2/f$ and applying an easily verified identity

$$\mathrm{re}[(1/f)(\partial/\partial t)f] = \tfrac{1}{2}(\partial/\partial t)(f\varepsilon)/(R^2 + f\varepsilon),$$

we obtain, by (16.1) with $w = f(s,t)$, after letting $|s| \to r+$,

$$(\partial/\partial t)\log[R^2(t) + f(s,t)\varepsilon(s,t)] = \tfrac{1}{2}\log R(t) \quad \text{for } |s| = r,\ 0 \leq t \leq T.$$

From the above it follows that $R^2(t) + f(s,t)\varepsilon(s,t) = e^{c(s)}R^2(t)$, where c does not depend on t. Hence, by the definition of ε,

$$|f(s,t)|^2 = e^{c(s)}R^2(t) \quad \text{for } |s| = r,\ 0 \leq t \leq T.$$

Now, taking into account the initial condition $f(s,0) = s$, we see that $c(s) = 0$ identically, and thus $|f(s,t)| = R(t)$ for $|s| = r$.

Since, by Theorem 5, any $f(\ ,t)$ is a homeomorphism, the above boundary relations imply that it maps Δ_r onto $\Delta_{R(t)}$. Finally, the relation $f(1,t) = 1$ is a consequence of (15.5) and $G(1,t) = 0$, where $G(w,t)$ denotes the right-hand side of (16.1). In this way the proof of Theorem 10 is completed.

The analogues of Theorems 7 and 8 can be proved exactly in the same way as the original theorems. Thus we obtain:

THEOREM 11. *Under the hypotheses of Theorem 9, if the partial derivative p_t of the dilatation (15.8) is positive a.e. in Δ_r and such that $\|p_t(\ ,t)\|_\infty \leq M < +\infty$ for $0 \leq t \leq T$, then the change of variable $t = T_0\tau$, where $T_0 = 1/M$, leads to an analogue of (16.1), where $\varphi(z,t)$ is replaced by $T_0\varphi(z, T_0\tau)$ and $\|T_0\varphi(\ , T_0\tau)\|_\infty \leq \tfrac{1}{2}$ for $0 \leq \tau \leq T/T_0$.*

THEOREM 12. *Under the hypotheses of Theorem* 10 *we have the estimates* (15.12) *and* (15.13) *for* $R(t) \le |w| \le 1$ *except perhaps for* w *such that* $|\varphi(w,t)| > \|\varphi(\ ,t)\|_\infty$ *for some* t.

17. Teichmüller mappings and quasiconformal mappings with invariant boundary points in the unit disc

Under a *Teichmüller mapping* of Δ onto itself we mean any qc mapping $f(\ ,t)$ whose complex dilatation $\varphi(\ ,t)$ has a.e. the form $t\bar{\varphi}/|\varphi|$, where $0 \le t < 1$ and φ is a function meromorphic in int Δ whose only singularities may be poles of the first order. In this section we consider only *normalized* Teichmüller mappings, i.e. mappings satisfying conditions (14.20) with fixed s_j, $j=1,2,3$, and suppose that $\varphi = \Phi'^2$, where Φ is holomorphic. Therefore we have a.e. in Δ

$$(17.1)\quad f_{\bar{s}}(s,t) = \mu(s,t) f_s(s,t),\ \ \mu(s,t) = t\,\overline{\Phi'(s)}/\Phi'(s),\ \ 0 \le t < 1.$$

We say that the mapping $f(\ ,t)$, uniquely determined (cf. Corollary 3, the classical theorems of Riemann and Osgood-Carathéodory, and the formula (9.1)) by the conditions (14.20) and (17.1), *corresponds* to Φ and t.

Investigation of the mappings in question was originated by Teichmüller [1, 2]. They play an important role in variational problems (cf. Sections 25 and 26 below). Simple examples show that normalized Teichmüller mappings of Δ onto itself keep, in general, the boundary points invariant. This interesting result is due to Reich and Strebel [2] and the proof is based on an ad hoc parametrical method leading to a parametrical equation valid at most for $t < \frac{1}{2}$, i.e. for Q-qc normalized Teichmüller mappings with Q bounded by some constant (depending on Φ) less than 3 (the final result concerning invariance is proved, however, for all Q). The general result, without the restriction $Q \le Q_0 < 3$ for the parametrical equation, has been obtained by Ławrynowicz [6] as a consequence of Theorems 5 and 6 together with Corollaries 7 and 8. The proof presented here (Ławrynowicz [8]) is still simpler. The result in question shows the power and beauty of the parametrical method. This result may be formulated as follows:

THEOREM 13. *Suppose that* Φ *is holomorphic in* int Δ, Φ' *has zeros at* s_j, $j=1,2,\dots$ (*we do not exclude the case where* $\Phi'(s) \ne 0$, $|s|<1$),

$$(17.2)\quad \iint_\Delta \frac{\overline{\Phi'(z)}}{\Phi'(z)}\,\frac{dxdy}{s-z} = 0 \ \ \textit{for}\ \ |s| = 1\ \ \textit{and}\ \ s = s_j,\ j=1,2,\dots$$

($x = \mathrm{re}\, z$, $y = \mathrm{im}\, z$), and

$$(17.3)\qquad \left| \frac{\Phi'(w) - \Psi_w(w,t)}{t\,\Phi'(w)} \right| \leq M, \quad 1 < M < +\infty, \quad \text{for } |w| < 1,$$

with

$$(17.4)\qquad \Psi(w,t) = \Phi(w) + t\Big[\,\overline{\Phi(w)} - \frac{1}{\pi}\Phi'(w) \iint_{\Delta} \frac{\overline{\Phi'(z)}}{\Phi'(z)} \frac{dxdy}{w-z}\Big], \quad 0 \leq t < 1.$$

Then there exists a unique solution $w = f(s,t)$ of

$$(17.5)\qquad w_t = \frac{1}{\pi} \iint_{\Delta} \frac{\varphi(z,t)}{w-z}\, dxdy, \quad \text{where } \varphi(w,t) = \frac{1}{1-t^2} \frac{\overline{\Psi_w(w,t)}}{\Psi_w(w,t)},$$

on $\Delta \times [0;1)$, subject to the initial condition $f(s,0) = s$, which for every t represents the Teichmüller mapping of Δ onto itself normalized by the conditions (14.20) and corresponds to Φ and t. Moreover, for every t we have

$$(17.6)\qquad \iint_{\Delta} \frac{\overline{\Psi_z(z,t)}}{\Psi_z(z,t)} \frac{dxdy}{w-z} = 0 \quad \text{for } |w| = 1 \text{ and } w = s_j,\ j = 1,2,\ldots,$$

and the mapping $f(\ ,t)$ satisfies the relations

$$(17.7)\qquad f(s,t) = s \quad \text{for } |s| = 1 \text{ and } s = s_j,\ j = 1,2,\ldots,$$

as well as

$$(17.8)\qquad \Psi(f(s,t),t) = \Phi(s) + t\,\overline{\Phi(s)} \quad \text{for } |s| < 1.$$

Remark 1. The orthogonality condition (17.2) is quite natural since it shows to be also necessary for Teichmüller mappings, as pointed out in Theorem 14.

Proof of Theorem 13. For greater clarity the proof is divided into six steps. In order to arrive at the most important relation (17.7) only Steps A-D are needed provided that we replace the differential equation in (17.5) by (14.21).

Step A. We begin with proving the existence of a unique solution $w = f(s,t)$ of (14.21) on $\Delta \times [0;1)$ with φ given in (17.5), subject to the initial condition $f(s,0) = s$, which for every t represents a qc. mapping of Δ onto itself, normalized by the conditions (14.20).

The change of variable $t = T_0\tau$ leads to an analogue of (14.21), where $\varphi(w,t)$ is replaced by

$$T_o\varphi(w, T_o\tau) = T_o(1 - T_o^2\tau^2)^{-1}\overline{\Psi_w(w, T_o\tau)}/\Psi_w(w, T_o\tau).$$

By Lemma 4 and formula (17.4), $\Psi(,t)$ is holomorphic in int Δ for any fixed t, so φ is measurable. Next we easily check that, given $\varepsilon > 0$, if we choose $T_o = \varepsilon(1 - \frac{1}{2}\varepsilon)$, then $\|T_o\varphi(, T_o\tau)\|_\infty \le \frac{1}{2}$ for $0 \le t \le 1 - \varepsilon$. In other words, if we take T_o, $T_o > 0$, sufficiently small, we can get t, $t < 1$, arbitrarily close to 1. Therefore, by Theorem 6 together with Corollary 8, the conclusion of Step A follows.

Step B. Let us observe next that since the solution $w = f(s,t)$ mentioned in Step A is unique, so is $\Psi(f(,t), t)$. We are going to try to guess $\Psi(f(,t), t)$. Relations (17.2) and (17.4) naturally suggest that $\Psi(f(,t), t)$ is perhaps given by (17.8). If so, then $f(,t)$ would be uniquely determined by (17.8) and (14.20). Indeed, by holomorphy of $\Psi(,t)$, the complex dilatation $\mu(,t)$ of $f(,t)$ clearly satisfies (17.1), so the required uniqueness would trivially follow from Corollary 3, the classical theorems of Riemann and Osgood-Carathéodory, and from the formula (9.1).

Step C. Now we have to verify that the qc mappings $w = f(s,t)$ of Δ onto itself, given by (17.8) and (14.20), really satisfy (14.21) with φ as in (17.5).

By Theorem 5 together with Corollary 7, $w = f(s,t)$ is a solution of the differential equation (14.21), where φ is defined by (14.2). Hence, by (17.1),

$$\varphi(w,t) = t^{-1}(1 - t^2)^{-1}\mu(\check{f}(w,t), t)\exp\{-2i \arg \check{f}_w(w,t)\}.$$

On the other hand, by (17.8) and holomorphy of $\Psi(,t)$, we have $\Psi_w(w,t) = \Phi'(\check{f}(w,t))\check{f}_w(w,t)$, whence, by (17.1) again,

$$\mu(\check{f}(w,t), t)\exp\{-2i \arg \check{f}_w(w,t)\} = t\,\overline{\Psi_w(w,t)}/\Psi_w(w,t).$$

Consequently we conclude that φ coincides with the function given in (17.5).

Step D. In order to check that also the differential equation in (17.5) is satisfied, we have to study more closely the boundary correspondence under $f(,t)$, so we check the formula (17.7). In fact this will be the most important result of this section.

By (17.8) we have $(d/dt)\,(f(s,t), t) = \Phi(s)$, $|s| < 1$, and

$$\Phi(s) = (1 - t^2)^{-1}[\Psi(f(s,t), t) - t\,\overline{\Psi(f(s,t), t)}], \quad |s| < 1.$$

Hence $w = f(s,t)$ satisfies a.e. in Δ the differential equation

(17.9) $\Psi_w(w,t)w_t + \Psi_t(w,t) = (1-t^2)^{-1}[\overline{\Psi(w,t)} - t\Psi(w,t)]$

i.e., in view of (17.4) and (17.2),

(17.10) $$\Psi_w(w,t)w_t = \frac{\Phi'(w)}{1-t^2}\Big[\frac{1}{\pi}\iint_\Delta \frac{\overline{\Phi'(z)}}{\Phi'(z)}\frac{dxdy}{w-z} - \frac{t}{\pi}\frac{\overline{\Phi'(w)}}{\Phi'(w)}\iint_\Delta \frac{\Phi'(z)}{\overline{\Phi'(z)}}\frac{dxdy}{\bar{w}-\bar{z}}\Big]$$

if $|w|<1$, $w\neq s_j$, and $w_t = 0$ if $|w|=1$ or $w=s_j$, $j=1,2,\ldots$

Therefore, by (17.4) again, (17.3), and the initial condition $f(s,0) = s$, we conclude that (17.7) certainly holds for $0\le t\le 1/M$ since

$$\left|\frac{\Psi_w(w,t)}{\Phi'(w)}\right| \ge 1 - t\left|\frac{\Phi'(w)-\Psi_w(w,t)}{t\,\Phi'(w)}\right| \ge 1 - tM$$

(note that the expression estimated by M does not depend on t). On the other hand, by (17.8) and (17.4), any function $f(s,\ )$, $|s|\le 1$, is real-analytic, so (17.7) holds in the whole interval $[0;1)$.

S t e p E. Now we are ready to verify the relation (17.6). Since the differential equation (14.21) may be written in the form (14.22), where $h(\ ,t)$ is holomorphic in Δ, then, by Lemma 4, the derivative $f'_{t\bar{s}}$ exists a.e. and, by Step C, we have a.e. in Δ

$$f'_{t\bar{s}} = (1-t^2)^{-1}\,\overline{\Psi_w(w,t)}/\Psi_w(w,t), \quad w = f(s,t).$$

We wish to utilize now the second of the Green's formulae (3.2) which, by (17.7), should give

$$\iint_\Delta \frac{\overline{\Psi_z(z,t)}}{\Psi_z(z,t)}\frac{dxdy}{w-z} = -\tfrac{1}{2}i(1-t^2)\int_{\partial\Delta} f_t(s,t)\Big|_{s=\check{f}(z,t)}\frac{dz}{w-z}.$$

Unfortunately, we have not proved (3.2) before, under sufficiently weak assumptions. Owing however to (17.7) we may continue $f_t(\ ,t)$ onto a larger disc $\Delta^{1+\varepsilon}$ by the value zero. Then we may drill in $\Delta^{1+\varepsilon}$ the disc $\Delta^\varepsilon(w)$ and continue the function h given by $h(z) = 1/(w-z)$, $z\in\Delta\setminus\operatorname{int}\Delta^\varepsilon(w)$, onto $D_\varepsilon(w) = \Delta^{1+\varepsilon}\setminus\operatorname{int}\Delta^{\frac{1}{2}\varepsilon}(w)$ so that it still be there of the class C^1 and have a compact support. Now, by the second of the formulae (3.1), we have

$$\iint_{D_\varepsilon(w)} \frac{\overline{\Psi_z(z,t)}}{\Psi_z(z,t)}h(z)dxdy = -(1-t^2)\iint_{D_\varepsilon(w)} f_t(s,t)\Big|_{s=\check{f}(z,t)} h_{\bar{z}}(z,t)$$
$$\times dxdy$$

for all sufficiently small $\varepsilon > 0$. The integral on the right-hand side has to be calculated separately on $\{z: |z| \leq 1, |z-w| > \varepsilon\}$, $\{z: |z| > 1, |z-w| > \varepsilon\}$, and $\{z: \frac{1}{2}\varepsilon \leq |z-w| \leq \varepsilon\}$.

The first integral vanishes since $f_t(s,t) = 0$ for $|s| \leq 1$. The second integral vanishes since h is holomorphic in the domain in question. Finally, the third integral tends to zero as $\varepsilon \to 0+$. Indeed, $|h_{\bar{z}}(z)|$ can be estimated by a constant multiplied by ε^{-2} and the area by $\frac{4}{3}\pi\varepsilon^2$. The function $f_t(\ ,t)$ is continuous since it differs from the function $G(\ ,t)$ considered in the proof of Theorem 6 (Step A) by a holomorphic function; cf. Corollary 7 and formula (14.22). By the relation (17.7) the function $f_t(\ ,t)$, restricted to the annulus in question, tends uniformly to zero as $\varepsilon \to 0+$, and this suffices to conclude (17.6).

S t e p <u>F</u>. It remains to show that <u>the equation</u> (14.21) <u>can be simplified to the form given in</u> (17.5). Writing (14.21) in the form (14.22), where $h(\ ,t)$ is holomorphic in int Δ, we observe, by (17.6) and the already proved formula for φ in (17.5), that $h(\ ,t)$ vanishes on fr Δ, and therefore it must vanish identically. This completes the proof.

Theorem 13 gives a sufficient condition for a qc mapping of Δ onto itself to keep the boundary points invariant. The converse problem is even simpler. We have the following result (relations (17.2) and (17.6) in this context are already due to Ahlfors [3], and relation (17.8) to Strebel [2]):

<u>THEOREM 14</u>. <u>Suppose that</u> Φ <u>is holomorphic in</u> int Δ <u>and</u> $f(\ ,t)$, $0 \leq t \leq 1$, <u>are Teichmüller mappings of</u> Δ <u>onto itself</u>, <u>normalized by the conditions</u> (14.20), <u>which correspond to</u> Φ <u>and</u> t. <u>Then</u> (17.7) <u>with</u> $|s_j| < 1$, $j = 1,2,\ldots$ (<u>we do not exclude the case where the set of points</u> s_j <u>is empty</u>), <u>implies</u> (17.2), (17.6), (17.5), <u>and</u> (17.8), <u>where</u> Ψ <u>is given by</u> (17.4).

<u>P r o o f</u>. Let us observe that since for $0 \leq t \leq 1$ there is exactly one qc mapping of Δ onto itself satisfying (17.1) and (14.20) (cf. Step B of the preceding proof), the same concerns $\Psi(f(\ ,t),t)$, where Ψ is given by (17.4). On the other hand, if $\Psi(f(\ ,t),t)$ were given by (17.8), then, by holomorphy of $\Psi(\ ,t)$ (cf. Lemma 4), f would satisfy (17.1); (14.20) being fulfilled by (17.7). Therefore $\Psi(f(\ ,t),t)$ is indeed given by (17.8). Now we can follow the argument given in Step C of the preceding proof to conclude that the qc mappings $w = f(s,t)$ of Δ onto itself satisfy (14.21) with φ as in (17.5). Next

the argument of Step E leads to (17.6) which for $t=0$ reduces to (17.2). Finally, by the argument given in Step F we conclude that the equation (14.21) can be simplified to the form given in (17.5), as desired.

Qc mappings which keep the boundary points invariant and, more generally, qc mappings with prescribed boundary values were extensively studied by Reich and Strebel [1-3], and R. S. Hamilton [1]. We will return to this subject in Section 26.

18. Extension to the case of an annulus

In analogy to the preceding section, under a normalized Teichmüller mapping of an annulus Δ_r onto $\Delta_{R(t)}$ we mean any qc mapping $f(\ ,t)$ satisfying $f(1,t)=1$ and (17.1) a.e. in Δ_r. We say that the mapping $f(\ ,t)$, uniquely determined (cf. Corollary 4) by the conditions $f(1,t)=1$ and (17.1) a.e. in Δ_r, corresponds to Φ and t. We are going to prove some analogues of Theorems 13 and 14 for an annulus (Ławrynowicz [7, 8]).

THEOREM 15. *Suppose that* Φ *is holomorphic in* $\operatorname{int}\Delta_r$, Φ' *has zeros at* s_j, $j=1,2,\ldots$ (*we do not exclude the case where* $\Phi'(s)\neq 0$, $r<|s|<1$),

(18.1) $F(s)=0$ *for* $|s|=1$ *and* $s=s_j$, $j=1,2,\ldots,$

$$(18.2)\quad F(s)=\frac{1}{\pi}\iint_{\Delta_r}\frac{\overline{\Phi'(z)}}{\Phi'(z)}\frac{dxdy}{s-z}$$

$$+\frac{1}{\pi}\iint_{\Delta_r}\sum_{n=1}^{+\infty}r^{4n}\Big[\frac{\overline{\Phi'(z)}/\Phi'(z)}{s-r^{2n}z}-\frac{1}{\bar{z}^3}\,\frac{\Phi'(z)/\overline{\Phi'(z)}}{r^{2n}-s\bar{z}}\Big]dxdy$$

($x=\operatorname{re} z$, $y=\operatorname{im} z$). *Let further* $\Phi(s)=\int(r^{4n}/s)\Phi'(r^{2n}/\bar{s})ds$, $r^{2n}<|s|<r^{2n-1}$, *be a holomorphic function, and let* $\Phi(s)=\Phi(s/r^{2n})$, $r^{2n+1}<|s|<r^{2n}$, *where* $n=1,2,\ldots$ *Finally, assume an estimate*

$$(18.3)\quad \left|\frac{\Phi'(w)-\Psi_w(w,t)}{t\,\Phi'(w)}\right|\leq M,\ 1<M<+\ ,\ \textit{for}\ r<|w|<1,$$

with Ψ *given by* (17.4). *Then the following statements hold.*

(i) *There exists a unique solution* $\rho=R(t)$ *of* (16.2) *with* φ *as in* (17.5) *on* $[0;1)$ *subject to the initial condition* $R(0)=r$.

(ii) *There exists a unique solution* $w=f(s,t)$ *of*

(18.4) $w_t = G(w,t)$,

(18.5) $$G(w,t) = \frac{1}{\pi}\iint\limits_{\Delta_{R(t)}} \frac{\varphi(z,t)}{w-z}\,dxdy$$

$$+ \frac{1}{\pi}\iint\limits_{R(t)} \sum_{n=1}^{+\infty} R^{4n}(t)\Big[\frac{\varphi(z,t)}{w-R^{2n}(t)z} - \frac{1}{\bar{z}^3}\,\frac{\overline{\varphi(z,t)}}{R^{2n}(t)-w\bar{z}}\Big]dxdy,$$

where $R^n(t) = [R(t)]^n$, *on* $\Delta \times [0;1)$, *subject to the initial condition* $f(s,0) = s$, *which for every* t *represents the Teichmüller mapping of* Δ_r *onto* $\Delta_{R(t)}$ *normalized by the condition* $f(1,t) = 1$ *and corresponds to* Φ *and* t.

(iii) *For every* t *we have*

(18.6) $G(w,t) = 0$ *for* $|w| = 1$ *and* $w = s_j$, $j = 1,2,\dots$,

and the mapping $f(\ ,t)$ *satisfies the relations*

(18.7) $f(s,t) = s$ *for* $|s| = 1$ *and* $s = s_j$, $j = 1,2,\dots$,

as well as

(18.8) $\Psi(f(s,t),t) = \Phi(s) + t\,\overline{\Phi(s)}$ *for* $r < |s| < 1$.

(iv) *If we assume, in addition, one of the cases*

(18.9) $$\operatorname{im}\Big\{\frac{1}{s}\,\frac{F(s) - t\,H(s)\,\overline{F(s)}}{1 - t\,H(s)\frac{\partial}{\partial s}[F(s)/H(s)]}\Big\} = C(t) \quad \textit{for} \quad |s| = R(t)$$

with

(18.10) $$\lim_{\substack{z\to s\\ \Phi'(z)\neq 0}} \frac{\overline{\Phi'(z)}}{\Phi'(z)} = H(s) \quad \textit{for} \quad r^{2n} < |s| < r^{2n-2},\ n = 1,2,\dots;$$

(18.11) $$\frac{\partial}{\partial s}\frac{F(s)}{H(s)} = -\frac{s}{\bar{s}}\,\frac{1+(\bar{s}/s)^2H^2(s)}{1-(\bar{s}/s)^2H^2(s)},\quad \frac{\overline{F(s)}}{F(s)} = -\frac{\bar{s}}{s} \quad \textit{for} \quad |s| = r$$

with

(18.12) $$\lim_{\substack{z\to s\\ \Phi'(z)\neq 0}} \frac{\overline{\Phi'(z)}}{\Phi'(z)} = H(s),\quad F(s) = 0 \quad \textit{for} \quad |s| = r;$$

and

(18.13) $F(s) = 0$ *for* $|s| = r$,

then we obtain correspondingly

(18.9′) $\Theta(s,t) = \vartheta(s) + \int_0^t (1-t^2)^{-1} c(t)dt$ for $|s| = r$,

(18.11′) $R(t) = r$ *for* $|s| = r$,

and

(18.13′) $f(s,t) = s$ *for* $|s| = r$,

where $(s) = \arg s$, $\Theta(s,t) = \arg f(s,t)$, *assuming* $\Theta(s,\ )$ *to be continuous for* $|s| = r$, $-\pi < \arg s \leq \pi$.

Proof. For greater clarity the proof is divided into seven steps.

Step *A. We begin with proving* (i). By the argument given in Step A of the proof of Theorem 13 (let us write, for short, Step 13A etc.) we can apply Theorem 10, by which (i) follows.

Step *B. The* same theorem implies *the existence of a unique solution* $w = f(s,t)$ *of* (16.1) *on* $\Delta_r \times [0;1)$ *with* φ *as in* (17.5), *subject to the initial condition* $f(s,0) = s$, *which for every* t *represents a qc mapping of* Δ_r *onto* $\Delta_{R(t)}$, *normalized by the condition* $f(1,t) = 1$. Then, as in Step 13B, *we conjecture* $\Psi(f(\ ,t), t)$ *to be given by* (18.8), what is supported by the fact that, by holomorphy of $\Psi(\ ,t)$ and Corollary 4, in this case $R(t)$ and $f(\ ,t)$ would be uniquely determined by (18.8) and $f(1,t) = 1$.

Step *C* Now, as in Step 13C, we verify that, owing to Theorem 9, *the qc mappings* $w = f(s,t)$ *of* Δ_r *onto* $\Delta_{R(t)}$, *given by* (18.8) *and* $f(1,t) = 1$, *really satisfy* (16.1) *with* φ *as in* (17.5).

Step D. Next *we check the formula* (18.7). By (18.8) we get, as in Step 13D, the differential equation (17.9), i.e., in view of (17.4) and (18.1),

(18.14) $$\Psi_w(w,t)w_t = \frac{\Phi'(w)}{1-t^2}\left[F(w) - t\frac{\overline{\Phi'(w)}}{\Phi'(w)}\overline{F(w)}\right] \text{ if } r < |w| < 1,\ w \neq s_j,$$

$$\text{and } w_t = 0 \text{ if } |w| = 1 \text{ or } w = s_j,\quad j = 1,2,\ldots$$

Arguing again as in Step 13D we conclude that, by (17.4), (18.3), (18.8), and the initial condition $f(s,0) = s$, (18.7) holds in the whole interval $[0;1)$.

Step *E*. Now we are ready to verify the relation (18.6). To this end we should like to proceed in analogy to Step 13E. Unfortunately the case $f(s,t)=s$ for $|s|=r$, $0\leq t<1$ is very restrictive, and we should like to consider the general case. Of course it is possible to calculate the integral

$$\int_{\partial\Delta_{R(t)}} f_t(s,t)\Big|_{s=\check{f}(z,t)} (w-z)^{-1}dz$$

by integration under the sign of integration with help of the method of shifting the contour $\partial\Delta_{R(t)}$ beyond the singular points (cf. Ławrynowicz [1], pp. 36-37), but we prefer to proceed in a more elementary way.

Let us continue quasiconformally the mappings $f(\ ,t)$, $0\leq t<1$, into the inner disc Δ_r by the formulae (16.3) and (16.4). Then the corresponding complex dilatations are given by (16.5) and (16.6). Consequently, the corresponding function φ^*, defined by (16.7), satisfies (16.9) and (16.10), but, by (17.1), this yields

(18.15) $\varphi^*(\ ,t)=\varphi(\ ,t)$ a.e., where φ is as in (17.5).

Therefore we may replace (16.1) by (14.22), where $h(\ ,t)$ is holomorphic in $\operatorname{int}\Delta$. Following now the procedure of Step 13E we arrive at the relation (17.6).

Let us observe next that, by (16.9), (16.10), and (18.15), we have

$$\iint_{\Delta} \frac{\varphi(z,t)}{w-z}dxdy = \iint_{\Delta_{R(t)}} \frac{\varphi(z,t)}{w-z}dxdy$$
$$+\sum_{n=1}^{+\infty}\Big[\iint_{\Delta(2n)} \frac{\varphi(z/R^{2n}(t),t)}{w-z}dxdy + \iint_{\Delta(2n-1)} e^{4i\arg z}\frac{\varphi(R^{2n}(t)/\bar{z},t)}{w-z}dxdy\Big],$$

where $\Delta(2n-1)$ and $\Delta(2n)$ are given by (16.11). Now the change of variables $z\mapsto R^{2n}(t)\bar{z}$ in the integral over $\Delta(2n-1)$ and $z\mapsto z/R^{2n}(t)$ in the integral over $\Delta(2n)$, $n=1,2,\ldots$, leads to the condition

$$\iint_{\Delta} \frac{\varphi(z,t)}{w-z}dxdy = \iint_{\Delta_{R(t)}} \frac{\varphi(z,t)}{w-z}dxdy \tag{18.16}$$
$$+\sum_{n=1}^{+\infty} R^{4n}(t)\iint_{\Delta_{R(t)}} \Big[\frac{\varphi(z,t)}{w-R^{2n}(t)z} - \frac{1}{\bar{z}^3}\,\frac{\varphi(z,t)}{R^{2n}(t)-w\bar{z}}\Big]dxdy.$$

It can be easily checked with help of the well known Weierstrass' test that the series of integrands in (18.6) is uniformly convergent in $\Delta_{R(t)}$. Since all integrals appearing in (18.6) are continuous in w and $\bar{w}$ (cf. the proof of Theorem 6, Step A), we can interchange the order of integration and summation in (18.6). Consequently, by (18.5), the definition of φ given in (17.5), and (17.6), the desired relation (18.6) follows.

S t e p F. In order to finish the proof of (ii) and (iii) it still remains to show that the equation (16.1) may be simplified to the form (18.4). Owing to Step E we already know that (16.1) may be replaced, by the continuation (16.3) and (16.4) of $w=f(s,t)$, with (14.22), where $h(\ ,t)$ is holomorphic in $\operatorname{int}\Delta$. Since, by (18.5), (18.6), and (18.16), $h(\ ,t)$ vanishes on $\operatorname{fr}\Delta$, it must vanish identically, and this completes the proof of (ii) and (iii).

S t e p G. We proceed to prove (iii). Given t, let $n(t)$ be the nonnegative integer chosen so that $r^{n(t)+1} < R(t) \le r^{n(t)}$. Hence the first relation in (18.14) remains valid if $r^{n+1} < |w| < r^n$, $n=1,\dots,n(t)-1$ (of course this set may be empty), and, in the case where $R(t) < r^{n(t)}$, also if $R(t) < |w| < r^{n(t)}$. By (17.4) this relation may be rewritten as

$$\{1 - t\frac{\overline{\Phi'(w)}}{\Phi'(w)}\frac{\partial}{\partial w}[\frac{\Phi'(w)}{\overline{\overline{\Phi'(w)}}}F(w)]\}w_t = \frac{1}{1-t^2}[F(w) - t\frac{\overline{\Phi'(w)}}{\Phi'(w)}\overline{F(w)}].$$

If we suppose (18.10) and t_0 denotes the smallest solution of the equation $R(t)=r^2$ in $[0;1)$ (if it has no solutions, we put $t_0=1$), then, by (18.3), for $0\le t<\min(t_0, 1/M)$ we have

$$(18.17)\quad w_t = \frac{1}{1-t^2}\,\frac{F(w) - t\,H(w)\,\overline{F(w)}}{1 - t\,H(w)\frac{\partial}{\partial w}[F(w)/H(w)]},\qquad \begin{array}{l} w=f(s,t),\ R(t)\le |w|<1,\\ |w|\ne r^{2n},\ 1\le n\le \tfrac{1}{2}n(t). \end{array}$$

Let us observe now that

$$f_t(s,t) = R(t)e^{i\Theta(s,t)}\{[1/R(t)]R'(t) + i\Theta_t(s,t)\}\quad\text{for } |s|=r.$$

Therefore, if we further assume (18.9), then, by the initial condition $f(s,0)=s$, (18.10) and (18.17) yield (18.9′) for $0\le t<\min(t_0, 1/M)$. On the other hand, by (18.8) and (17.4), any function $f(s,\ )$, $r\le |s| \le 1$, is real-analytic, so (18.9′) holds in the whole interval $[0;1)$. Similarly, in the case of (18.11) and (18.12) we arrive at

$$(18.18)\quad \operatorname{re}\{\tfrac{1}{s}[F(s) - t\,H(s)F(s)]/[1 - t\,H(s)\tfrac{\partial}{\partial s}[F(s)/H(s)]]\} = 0\quad\text{for } |s|=r$$

and, consequently, (18.17) yields (18.11′). Finally, in the case of (18.13) we obtain (18.9) with $C(t)=0$ and $R(t)$ replaced by r, and (18.18). Hence, by (18.17), the relation (18.13′) follows. In this way also the assertion (iii) is proved.

COROLLARY 10. *Suppose that* Φ *is holomorphic in* $\operatorname{int}\Delta_r$, Φ' *has zeros at* s_j, $j=1,2,\ldots$ (*we do not exclude the case where* $\Phi'(s)\neq 0$, $r<|s|<1$),

$$(18.19)\qquad \iint_{\Delta_r} \frac{\overline{\Phi'(z)}}{\Phi'(z)}\,\frac{dxdy}{s-z} = 0 \quad \text{for}\quad |s|=1,\ |s|=r,\ \text{and}\ \ s=s_j,\ j=1,2,\ldots,$$

and an estimate (18.3) *holds with* Ψ *given by* (17.4). *Then there exists a unique solution* $w=f(s,t)$ *of*

$$(18.20)\qquad w_t = \frac{1}{\pi}\iint_{\Delta_r}\frac{\varphi(z,t)}{w-z}\,dxdy, \quad \text{where}\quad \varphi(w,t)=\frac{1}{1-t^2}\,\frac{\overline{\Psi_w(w,t)}}{\Psi_w(w,t)},$$

on $\Delta_r \times [0;1)$, *subject to the initial condition* $f(s,0)=s$, *which for every* t *represents the Teichmüller mapping of* Δ_r *onto itself normalized by the condition* $f(1,t)=1$ *and corresponds to* Φ *and* t. *Moreover, for every* t *we have*

$$(18.21)\qquad \iint_{\Delta_r}\frac{\overline{\Psi_z(z,t)}}{\Psi_z(z,t)}\,\frac{dxdy}{w-z}=0 \quad \text{for}\quad |w|=1,\ |w|=r,\ \text{and}\ \ w=s_j,\ j=1,2,\ldots,$$

and the mapping $f(\ ,t)$ *satisfies the relations*

$$(18.22)\qquad f(s,t)=s \quad \text{for}\quad |s|=1,\ |s|=r,\ \text{and}\ \ s=s_j,\ j=1,2,\ldots,$$

as well as (18.8).

Proof. The corollary is an immediate consequence of Theorem 15 in the case (18.13). In fact, in this case (18.4) and (18.5) with φ as in (17.5) reduce to the form (18.20) since any function

$$G(w,t)-\frac{1}{\pi}\iint_{\Delta_{R(t)}}\frac{\varphi(z,t)}{w-z}\,dxdy,\quad R(t)<|w|<1,$$

being holomorphic in $\operatorname{int}\Delta_{R(t)}$ and vanishing on $\operatorname{fr}\Delta_{R(t)}$, must vanish identically.

Actually Corollary 10 is the most direct and natural analogue of Theorem 13 for an annulus and its direct proof would be much simpler than that of Theorem 15. The way we have chosen is motivated by the fact that the class of mappings covered by Corollary 10 is rather small.

THEOREM 16. Suppose that Φ is holomorphic in $\operatorname{int}\Delta_r$, and $f(\ ,t)$, $0 \leq t \leq 1$, are Teichmüller mappings of Δ_r onto $\Delta_{R(t)}$, normalized by the condition $f(1,t) = 1$, which correspond to Φ and t. Then (18.7) with $r < |s_j| < 1$, $j = 1,2,\ldots$ (we do not exclude the case where the set of points s_j is empty), implies (18.1), (18.6), (18.4), and (18.8), where F, G, and Ψ are given by (18.2), (18.5), and (17.4), respectively. If we assume, in addition, one of the cases (18.9′) with (18.10), (18.11′) with (18.12), and (18.13′), then we obtain correspondingly (18.9), (18.11), and (18.13).

Proof. Except for the last sentence, the theorem may be proved along the same lines as Theorem 14. The truth of the last sentence can be easily checked directly by taking into account the relation (18.17).

Let us finally notice the following immediate

COROLLARY 11. Suppose that Φ is holomorphic in $\operatorname{int}\Delta_r$, and $f(\ ,t)$, $0 \leq t < 1$, are Teichmüller mappings of Δ_r onto $\Delta_{R(t)}$, normalized by the condition $f(1,t) = 1$, which correspond to Φ and t. Then (18.22) with $r < |s_j| < 1$, $j = 1,2,\ldots$ (we do not exclude the case where the set of points s_j is empty), implies (18.19), (18.21), (18.20), and (18.8), where Ψ is given by (17.4).

Normalized Teichmüller mappings of an annulus Δ_r onto $\Delta_{R(t)}$ have also been investigated by Zając [1, 2]. Extending a result of Ahlfors [3] he proved the following

LEMMA 17. Suppose that $f[\varkappa](\ ,t)$ is the qc mapping of Δ_r onto $\Delta_{R[\varkappa](t)}$, uniquely determined by the conditions $f[\varkappa](1,t) = 1$ and

$$\frac{\partial}{\partial \bar{s}} f[\varkappa](s,t) = t\,\varkappa(s)\,\frac{\partial}{\partial s} f[\varkappa](s,t) \quad \text{a.e. in } \Delta_r,\ 0 \leq t \leq T,\ T < 1,$$

and consider also the corresponding mapping $f[i\ ](\ ,t)$. Let

$$(18.23)\qquad g[\varkappa] = \dot{f}[\varkappa] + i\,\dot{f}[i\varkappa], \quad \dot{f}[\varkappa] = (\partial/\partial t) f[\varkappa](\ ,0).$$

Then

$$(18.24)\qquad g[\varkappa](s) = 0 \quad \text{for } r \leq |s| \leq 1$$

iff

$$(18.25)\qquad \dot{f}[\varkappa](s) = 0 \quad \text{for } |s| = 1$$

and

(18.26) $\dot{f}[\varkappa](s) = (s/\pi)\iint_{\Delta_r}(1/z^2)\varkappa(z)dxdy$ for $|s| = r$.

Proof. By the identity

$$\mathrm{re}\{\bar{s}\,\dot{f}[\varkappa](s)\} = \tfrac{1}{2}\lim_{t\to 0+}\frac{\{f[\varkappa](s,t)-s\}^2}{ts\,f[\varkappa](s,t)},\quad |s| = 1,$$

which yields $\mathrm{re}\{\bar{s}\,\dot{f}[\varkappa](s)\} = 0$, we have there

(18.27) $\mathrm{re}\{\bar{s}\,g[\varkappa](s)\} = -\mathrm{im}\{\bar{s}\,\dot{f}[i\varkappa](s)\}$,

(18.28) $\mathrm{im}\{\bar{s}\,g[\varkappa](s)\} = \mathrm{im}\{\bar{s}\,\dot{f}[i\varkappa](s)\}$,

and

(18.29) $\bar{s}\,\dot{f}[\varkappa](s) = i\,\mathrm{im}\{\bar{s}\,g[\varkappa](s)\}$, $|s| = 1$.

Similarly, by the identity

$$\mathrm{re}\{\bar{s}\,\dot{f}[\varkappa](s)\} = \tfrac{1}{2}\lim_{t\to 0+}\frac{\{f[\varkappa](s,t)-s\}^2}{ts\,f[\varkappa](s,t)} + r\frac{d}{dt}R[\varkappa](0),\quad |s| = r,$$

which yields $\mathrm{re}\{\bar{s}\,\dot{f}[\ ](s)\} = r(d/dt)R[\ ](0)$, we have there

(18.30) $\mathrm{re}\{\bar{s}\,g[\varkappa](s)\} = r(d/dt)R[\varkappa](0) - \mathrm{im}\{\bar{s}\,\dot{f}[i\varkappa](s)\}$,

(18.31) $\mathrm{im}\{\bar{s}\,g[\varkappa](s)\} = r(d/dt)R[i\varkappa](0) + \mathrm{im}\{\bar{s}\,\dot{f}[i\varkappa](s)\}$,

and

(18.32) $\bar{s}\,\dot{f}[\varkappa](s) = r\dfrac{d}{dt}\{R[\varkappa](0) - i\,R[i\varkappa](0)\} + i\,\mathrm{im}\{\bar{s}\,g[\varkappa](s)\}$, $|s| = r$.

Moreover, by Theorem 9, $\rho = R[\varkappa](t)$ and $\rho = R[i\varkappa](t)$ satisfy the differential equation (16.2) with

$$\varphi(w,t) = \frac{1}{1-t^2}\varkappa(\check{f}[\varkappa](w,t))\exp\{-2i\arg\frac{\partial}{\partial w}\check{f}[\varkappa](w,t)\}$$

and

$$\varphi(w,t) = \frac{1}{1-t^2}\varkappa(\check{f}[i\varkappa](w,t))\exp\{-2i\arg\frac{\partial}{\partial w}\check{f}[i\varkappa](w,t)\},$$

respectively, so we get

(18.33) $$\frac{d}{dt}R[\varkappa](0) = \frac{r}{2\pi}\iint_{\Delta_r}[\frac{\varkappa(z)}{z^2} + \frac{\overline{\varkappa(z)}}{\bar{z}^2}]dxdy$$

and

$$(18.34)\quad \frac{d}{dt}R[i\varkappa](0) = \frac{ir}{2\pi}\iint_{\Delta_r}\left[\frac{\varkappa(z)}{z^2} - \frac{\overline{\varkappa(z)}}{\bar{z}^2}\right]dxdy.$$

Therefore, if (18.24) holds, then (18.25) is an immediate consequence of (18.29), and (18.26) – of (18.32)-(18.34). Conversely, if (18.25) and (18.26) hold, then, by (18.29) and (18.32)-(18.34), the function h^* of s with values $s^{-1}g[\varkappa](s)$ is real-valued on $\operatorname{fr}\Delta_r$. On the other hand, by Theorem 9, $w = f(s,t)$ is a solution of the differential equation (16.1) in $\Delta_r \times [0;T]$. Since this equation can be written in the form (14.22), where $h(\ ,t)$ is holomorphic in $\operatorname{int}\Delta_r$, then, by Lemma 4, the derivative f'_{ts} exists a.e. and, by (18.23), we have $g_{\bar{s}}[\varkappa](s) = 0$ a.e. in Δ_r. Hence, by the same lemma, $g[\varkappa]|\operatorname{int}\Delta_r$ is holomorphic and, by (18.27), (18.28), (18.30), and (18.31), $g[\varkappa]$ is continuous (in the <u>closed</u> annulus Δ_r). The function h^* has the same properties and is also real-valued on $\operatorname{fr}\Delta_r$. Consequently h^* is a constant function and, by $g[\varkappa](1) = 0$, we conclude (18.24).

Now it seems that, owing to Lemma 17, we can get another natural analogue of Theorem 13 for an annulus if instead of the conditions (18.9)-(18.13) we suppose that

$$(18.35)\quad F(s) = \frac{1}{\pi}\iint_{\Delta_r}\frac{\overline{\Phi'(z)}}{\Phi'(z)}\frac{dxdy}{z^2}s \quad \text{for } |s| = r.$$

Pursuing this idea Zając [1] obtained the following result which, in our context, is a corollary:

<u>COROLLARY 12</u>. <u>If we replace in Theorem</u> 15 <u>the conditions</u> (18.9) <u>and</u> (18.10) <u>by</u> (18.35), <u>then the formulae</u> (18.2) <u>and</u> (18.5) <u>reduce to</u>

$$(18.36)\quad F(s) = \frac{1}{\pi}\iint_{\Delta_r}\frac{\overline{\Phi'(z)}}{\Phi'(z)}\frac{dxdy}{s-z} + \frac{1}{\pi}\iint_{\Delta_r}\sum_{n=1}^{+\infty}\frac{\overline{\Phi'(z)}}{\Phi'(z)}\frac{r^{4n}}{s-r^{2n}z}dxdy$$

<u>and</u>

$$(18.37)\quad G(w,t) = \frac{1}{\pi}\iint_{\Delta_{R(t)}}\frac{\varphi(z,t)}{w-z}dxdy + \frac{1}{\pi}\iint_{\Delta_{R(t)}}\sum_{n=1}^{+\infty}\varphi(z,t)\frac{R^{4n}(t)}{w-R^{2n}(t)z}dxdy,$$

<u>respectively</u>. <u>If</u>, <u>in addition</u>, (18.9) <u>and</u> (18.10) <u>hold then we obtain</u> (18.9′) <u>and</u>

$$(18.38)\quad C(0) = \frac{1}{r}\operatorname{im}F(r).$$

If we complete in Theorem 16 the conditions (18.9′) and (18.10) by (18.38), then the formulae (18.2) and (18.5) reduce to (18.36) and (18.37), respectively, and we obtain (18.35) as well as (18.38) for every t.

Proof. The first sentence is a direct consequence of Lemma 17. The second sentence may be verified as in Step 15G. The truth of the third sentence may be checked directly by taking into account the relation (18.17).

All results of this section have their natural analogues in the case $r=0$, i.e. for mappings discussed in the preceding section with an additional invariant point 0.

19. Line distortion under quasiconformal mappings

We begin with giving a simple example of application of Theorem 5 for the class E_Q of Q-qc mappings with the radial symmetry: A function f is said to be of the class E_Q if it belongs to S_Q and if $f(s)=e^{i\arg s}f(|s|)$ for $0<|s|<1$. This class has been introduced and investigated by Ławrynowicz [4] (for Q-qc mappings of this class with conformal extension cf. Kühnau [9]). The class E_Q and analogous classes $E_Q^{r,R}$ of mappings f which belong to $S_Q^{r,R}$ and fulfil the condition $f(s)=e^{i\arg s}f(|s|)$ for $r\leq|s|\leq 1$ are of basic importance in electrical engineering (cf. Section 27 below). Clearly, $E_Q^{0,0}=E_Q$.

LEMMA 18. A function f belongs to the class E_Q iff one of the following equivalent conditions holds:

(i) $f=f^*|\Delta$, where $f^*\colon \mathbb{E}\to\mathbb{E}$ is Q-qc, $f^*(0)=0$, $f^*(1)=1$, $f^*(\infty)=\infty$,

$$f^*(s)=e^{-ia}/\overline{f^*(e^{ia}/\bar{s})}\quad \text{for}\quad s\neq 0,\infty.$$

and a is a real number such that a/π is irrational;

(ii) $f=f^*|\Delta$, where $f^*\colon \mathbb{E}\to\mathbb{E}$ is a Q-qc mapping whose complex dilatation μ^* satisfies the conditions

$$(19.1)\qquad \mu^*(s)=e^{2ia+4i\arg s}\,\overline{\mu^*(e^{ia}/\bar{s})}\quad \text{a.e. in}\ \mathbb{C}$$

and $f^*(0)=0$, $f^*(1)=1$, $f^*(\infty)=\infty$;

(iii) f belongs to S_Q and its complex dilatation μ satisfies $\mu(s)=e^{2i\arg s}\mu(|s|)$ a.e. in Δ;

(iv) f belongs to S_Q and $s f_s(s)-\bar{s}f_{\bar{s}}(s)=f(s)$ a.e. in Δ;

(v) $f(s) = \exp(-\int_{|s|}^{1} \frac{1+\mu(q)}{1-\mu(q)} \frac{dq}{q} + i \arg s)$ *for* $0<|s|\le 1$,

and $f(s)=0$ *for* $s=0$, *where* μ *is measurable*, $|\mu(s)|<1$ and $\|\mu\|_\infty \le (Q-1)/(Q+1)$.

An analogous result is valid for any class $S_Q^{r,R}$.

Proof. Since f^*, defined by $f^*(s)=f(s)$ for $|s|\le 1$ and by $f^*(s) = 1/\overline{f(1/\bar{s})}$ for $1<|s|<+\infty$, $f^*(\infty)=\infty$, where $f\in E_Q$, maps $\mathbb{E}$ onto itself Q-quasiconformally with $f^*(0)=0$, $f^*(1)=1$, and satisfies

$$e^{-ia}/\overline{f(e^{ia}/\bar{s})} = e^{-ia}f^*(s/e^{-ia}) = e^{-ia}e^{ia+i\arg s}f^*(|s|) = f^*(s)$$

for $0<|s|<+\infty$ and arbitrary real a, then (i) follows. Conversely, if (i) holds, then it is sufficient to prove that $|f(s)|=1$ for $|s|=1$, and so we only have to verify that $f(s)=e^{i\arg s}f(|s|)$ for $0<|s|\le 1$. To this end we notice first that setting $z=e^{ia}/\bar{s}$ in the equation

$$f^*(z) = e^{-ia}/\overline{f^*(e^{ia}/\bar{z})}$$

we get

$$f^*(e^{ia}/\bar{s}) = e^{-ia}/\overline{f^*(e^{2ia}s)},\ \text{i.e.}\ \ 1/\overline{f^*(e^{ia}/\bar{s})} = e^{-ia}f^*(e^{2ia}s).$$

Hence

$$f^*(s) = e^{-ia}/\overline{f^*(e^{ia}/\bar{s})} = e^{-2ia}f^*(e^{2ia}s).$$

Suppose now that we have

$$f^*(s) = e^{-2(n-1)ia}f^*(e^{2(n-1)ia}s)$$

for a positive integer n. Taking $z=e^{2(n-1)ia}s$ in $f^*(z)=e^{-2ia}\times f^*(e^{2ia}z)$, we obtain

$$f^*(e^{2(n-1)ia}s) = e^{-2ia}f^*(e^{2nia}s).$$

Consequently

$$f^*(s) = e^{-2(n-1)ia}f^*(e^{2(n-1)ia}s) = e^{-2nia}f^*(e^{2nia}s). \tag{19.2}$$

Thus (19.2) holds for any positive integer n, and it can easily be seen that it holds for any integer n. Let us observe that since a/π is irrational, by a theorem of Kronecker, for any real ϑ there exists a sequence $(2n_k a + 2m_k\pi)$, where m_k, n_k, $k=1,2,\dots$, are integers, tending to ϑ as $k\to+\infty$. Setting $\vartheta = -\arg s$ we obtain $f^*(s) = e^{i\arg s}\times f^*(|s|)$, and since $f^*|\Delta = f$, the desired condition follows.

Condition (i) implies, by Lemma 1, that f_s, $f_{\bar{s}}$ exist a.e. in Δ and that, by (1.2),

$$f^*_s(s) = \frac{-e^{-ia}}{\{\overline{f^*(e^{ia}/\bar{s})}\}^2}\,\frac{\partial}{\partial s}\overline{f^*(e^{ia}/\bar{s})} = \frac{-e^{-ia}}{\{\overline{f^*(e^{ia}/\bar{s})}\}^2}\,\overline{f^*_{\bar{s}}(e^{ia}/\bar{s})}$$

$$= \frac{-e^{-ia}}{\{\overline{f^*(e^{ia}/\bar{s})}\}^2}\,[\overline{f^*_z(z)}]_{z=e^{ia}/\bar{s}}\,\overline{\frac{\partial}{\partial\bar{s}}(e^{ia}/\bar{s})} = \frac{1}{s^2\{\overline{f^*(e^{ia}/\bar{s})}\}^2}\,[\overline{f^*(z)}]_{z=e^{ia}/\bar{s}}$$

and, similarly,

$$f^*_{\bar{s}}(s) = \frac{-e^{-ia}}{\{\overline{f^*(e^{ia}/\bar{s})}\}^2}\,[\overline{f^*_{\bar{z}}(z)}]_{z=e^{ia}/\bar{s}}\,\overline{\frac{\partial}{\partial\bar{s}}(e^{-ia}/\bar{s})}$$

$$= \frac{1}{\bar{s}^2\{\overline{f^*(e^{ia}/\bar{s})}\}^2}\,[\overline{f^*_{\bar{z}}(z)}]_{z=e^{ia}/\bar{s}}\,.$$

Hence, denoting by μ^* the complex dilatation of f^*, we get for μ^* the condition (19.1) required in (ii). Conversely, suppose (ii). By Corollary 3 the function f^* is determined uniquely. On the other hand, the function f^{**}, given by

$$f^{**}(s) = e^{ia}/\overline{f^*(e^{ia}/\bar{s})},\ s \neq 0, \infty;\ f^{**}(s) = s,\ s = 0, \infty,$$

is also defined in $\mathbb{E}$ and maps it onto itself Q-quasiconformally with $f^{**}(0) = 0$, $f^{**}(1) = 1$, $f^{**}(\infty) = \infty$, and its complex dilatation μ^{**} satisfies

$$\mu^{**}(s) = e^{2ia + 4i\arg s}\,\overline{\mu^*(e^{ia}/\bar{s})}\quad \text{a.e. in } \mathbb{C}$$

Since $\mu^{**}(s) = \mu^*(s)$ a.e., then $f^{**} = f^*$ which proves that the condition (ii) is also sufficient.

Next we are going to prove (iii). By Lemma 1, if $f \in E_Q$, then f_s, $f_{\bar{s}}$ exist a.e. in Δ and, with the notation

$$r = |s|,\ \varepsilon = e^{i\arg s},\ \hat{f}(r) = [f(s)]_{s=r},$$

we have

$$f_s(s) = \frac{\partial}{\partial r}\{\varepsilon\,\hat{f}(r)\}\,\frac{\partial}{\partial s}(s\bar{s})^{\frac{1}{2}} + \frac{\partial}{\partial\varepsilon}\{\varepsilon\,\hat{f}(r)\}\,\frac{\partial}{\partial\varepsilon}(s/\bar{s})^{\frac{1}{2}}$$

$$= \tfrac{1}{2}(\bar{s}/s)^{\frac{1}{2}}\,\varepsilon\,\hat{f}'(r) + \tfrac{1}{2}(s\bar{s})^{-\frac{1}{2}}\,\hat{f}(r).$$

Thus

(19.3) $f_s(s) = \frac{1}{2}\{[\hat{f}'(r)]_{r=|s|} + (1/|s|)f(|s|)\}$,

and, similarly,

(19.4) $f_{\bar{s}}(s) = \frac{1}{2}e^{2i\arg s}\{[\hat{f}'(r)]_{r=|s|} - (1/|s|)f(|s|)\}$.

Hence $\mu(s) = e^{2i\arg s}\mu(|s|)$ a.e. in Λ, i.e. we arrive at (iii). Suppose now (iii). Setting $\mu^*(s) = \mu(s)$ for $|s| \leq 1$ and $\mu^*(s) = e^{4i\arg s} \times \overline{\mu(1/\bar{s})}$ for $1 < |s| < +\infty$, we see that

$$\exp(2ia + 4i\arg s)\overline{\mu^*(e^{ia}/\bar{s})}$$

$$= \exp[-2ia + 4i\arg(s/e^{-ia})]\overline{\mu^*(e^{ia}/\bar{s})} = \exp(-2ia)\mu^*(e^{ia}s)$$

$$= \exp(-2ia)\exp(2ia + 2i\arg s)\mu^*(|s|) = \mu^*(s)$$

for any real a. Hence we conclude (ii), where f^* is defined by $f^*(s) = f(s)$ for $|s| \leq 1$ and by $f^*(s) = 1/\overline{f(1/\bar{s})}$ for $1 < |s| < +\infty$, $f^*(\infty) = \infty$.

In turn we proceed to verify (iv). Starting from $f \in E_Q$ we have, as before, the existence of f_s, $f_{\bar{s}}$ a.e. and the relations (19.3) and (19.4). Hence $sf_s(s) - \bar{s}f_{\bar{s}}(s) = e^{i\arg s}f(|s|)$ a.e. in Λ, as desired. Conversely, if (iv) holds, then, with the notation $r = |s|$ and $\varepsilon = e^{i\arg s}$, we have

$$\frac{\partial}{\partial\varepsilon}\{\frac{1}{\varepsilon}f(r\varepsilon)\} = -\frac{1}{\varepsilon^2}f(r\varepsilon) + \frac{1}{\varepsilon}\{f_s(s)\frac{\partial}{\partial\varepsilon}(r\varepsilon) + f_{\bar{s}}(s)\frac{\partial}{\partial\varepsilon}(r/\varepsilon)\}$$

$$= (1/\varepsilon^2)\{r\varepsilon f_s(s) - (r/\varepsilon)f_{\bar{s}}(s) - f(r\varepsilon)\}$$

$$= e^{-2i\arg s}\{s f_s(s) - \bar{s} f_{\bar{s}}(s) - f(s)\} = 0 \quad \text{a.e. in } \Lambda.$$

Consequently $e^{-i\arg s}f(s) = C(|s|)$, and choosing $s \geq 0$ we obtain $C(|s|) = f(|s|)$. Thus $f \in E_Q$.

Finally we prove (v). Starting from $f \in E_Q$ we have, as before, the existence of f_s, $f_{\bar{s}}$ a.e. and the relations (19.3) and (19.4). Hence, by (iii),

$$\mu(r) = e^{-2i\arg s}\frac{f_{\bar{s}}(s)}{f_s(s)} = \frac{\hat{f}'(r) - (1/r)\hat{f}(r)}{\hat{f}'(r) + (1/r)\hat{f}(r)}, \quad r = |s|, \text{ a.e. in } \Lambda.$$

Consequently, $\dfrac{\hat{f}'(r)}{\hat{f}(r)} = \dfrac{1+\mu(r)}{1-\mu(r)}(1/r)$ a.e. in Λ and, by $\hat{f}(1) = f(1) = 1$,

$$\hat{f}(|s|) = \exp\left(-\int_{|s|}^{1} \frac{1+\mu(q)}{1-\mu(q)} \frac{dq}{q}\right),$$

where the integral exists by the relation $f \in E_Q$. Since this relation and $\hat{f}(r) = [f(s)]_{s=r}$, $0 < r \le 1$, imply $f(s) = e^{i \arg s}\hat{f}(|s|)$ for $0 < |s| \le 1$ and $f(0) = 0$, we conclude that (v) holds indeed. Conversely, if (v) holds, then

$$\frac{\hat{f}'(r) - (1/r)\hat{f}(r)}{\hat{f}'(r) + (1/r)\hat{f}(r)} = \mu(r), \; r = |s|, \text{ a.e. in } \Delta, \quad \text{where } \hat{f}(r) = [f(s)]_{s=r}.$$

On the other hand, (v) implies $f(s) = e^{i \arg s} f(|s|)$ for $0 < |s| \le 1$; hence (cf. (19.3) and (19.4)) we obtain $f_s(s) = \frac{1}{2}\{\hat{f}'(r) - (1/r)\hat{f}(r)\}$ and $f_{\bar{s}}(s) = \frac{1}{2}e^{2i \arg s}\{\hat{f}'(r) - (1/r)\hat{f}(r)\}$, $r = |s|$, a.e. in Δ. Therefore $f_{\bar{s}}(s) = e^{2i \arg s}\mu(|s|)f_s(s)$ a.e. in Δ. Now, let us consider the Beltrami equation $w_{\bar{s}} = e^{2i \arg s}\mu(|s|)w_s$ a.e. in Δ. By Corollary 3 there exists exactly one mapping f^* of the class S_Q which has μ^*, $\mu^*(s) = e^{2i \arg s}\mu(|s|)$, as its complex dilatation a.e. in Δ. Hence (iii) follows. Since the condition (iii) is necessary and sufficient for $f \in E_Q$, we can repeat the previous considerations which show that if $f \in E_Q$ then we get (v) with f^* substituted for f. Hence $f^* = f$. Since $f^* \in E_Q$, then also $f \in E_Q$, and the proof is completed.

In the sequel we only need conditions (iii) and (v). In view of (iii) Theorem 5 yields (Ławrynowicz [4]):

COROLLARY 13. *$|f(s) - s|/\log Q < 1/e \doteq 0.37$ for $f \in E_Q$, $1 < Q < +\infty$, and $|s| \le 1$. The constant $1/e$ is the best possible.*

P r o o f. Let us use the notation of Theorem 5. If we take in it, in particular, $\mu(s,t) = t\mu(s)$, $|s| \le 1$, $0 \le t \le 1$, then (14.2) becomes

$$\varphi(w,t) = \{1/t(1 - |\mu(\check{f}(w,t), t)|^2)\}\mu(\check{f}(w,t), t)\exp\{-2i \arg \check{f}_w(w,t)\}.$$

On the other hand, since the complex dilatation of id_Δ is 0, then, by (9.1), we have

$$(19.5) \quad \mu(\check{f}(w,t), t) = -\mu_*(w,t)\exp\{2i \arg \check{f}_w(w,t)\},$$

where $\mu_*(\ ,t)$ denotes the complex dilatation of $\check{f}(\ ,t)$. Consequently, $\varphi(w,t) = -(1/t)\mu_*(w,t)/(1 - |\mu_*(w,t)|^2)$. Let us apply now Theorem 5 with $f(s,1) = f(s)$, $|s| \le 1$, and insert the above expression for φ in (14.1). Then we can rearrange this differential equation as follows:

$$w_t = \frac{w(1-w)}{\pi}\iint_\Delta \frac{-1}{1-|\mu_*(z,t)|^2}\left[\frac{\mu_*(z,t)}{z(1-z)(w-z)} + \frac{\overline{\mu_*(z,t)}}{\bar z(1-\bar z)(1-w\bar z)}\right]dxdy$$

$$= -\frac{w(1-w)}{\pi t}\iint_{\mathbb{C}} \frac{\hat\mu(z,t)}{1-|\hat\mu(z,t)|^2}\,\frac{dxdy}{z(1-z)(w-z)},$$

where $\hat\mu(w,t) = \mu_*(w,t)$ for $|w| \le 1$ and $\hat\mu(w,t) = e^{4i\arg w}\overline{\mu_*(1/\bar w,\, t)}$ for $|w| > 1$.

By Lemma 18, in particular condition (iii), the relation $f(\ ,1) \in E_Q$ implies $f(\ ,t) \in E_{Q(t)}$, $0 \le t \le 1$, so, in the polar coordinates $|z| = \rho$, $\arg w = \vartheta$, we get

$$w_t = -\frac{w(1-w)}{\pi t}\int_0^{+\infty}\int_{-\pi}^{\pi} \frac{e^{2i\vartheta}\hat\mu(\rho,t)}{1-|\hat\mu(\rho,t)|^2}\,\frac{\rho d\vartheta}{\rho e^{i\vartheta}(1-\rho e^{i\vartheta})(w-\rho e^{i\vartheta})}d\rho$$

$$= -\frac{w(1-w)}{\pi i t}\int_0^{+\infty} \frac{(1/\rho)\hat\mu(\rho,t)}{1-|\hat\mu(\rho,t)|^2}\int_{\partial\Delta\rho} \frac{dz}{(1-z)(w-z)}d\rho.$$

By the theorem of residues

$$(19.6)\qquad \int_{\partial\Delta\rho} \frac{dz}{(z-1)(z-w)} = \begin{cases} 0 & \text{for } 0 < \rho < |w|, \\ 2\pi i/(w-1) & \text{for } |w| < \rho < 1, \\ 0 & \text{for } 1 < \rho < +\infty. \end{cases}$$

Consequently,

$$w_t = 2(w/t)\int_{|w|}^{1} (1/\rho)\hat\mu(\rho,t)(1-|\hat\mu(\rho,t)|^2)^{-1}d\rho.$$

Therefore, since $\hat\mu(\rho,t) = \mu_*(\rho,t)$ for $0 \le \rho \le 1$, $0 \le t \le 1$, we finally get the following <u>parametrical</u> <u>equation</u> for $f(\ ,t) \in E_{Q(t)}$:

$$(19.7)\qquad w_t = 2(w/t)\int_{|w|}^{1} (1/\rho)\mu_*(\rho,t)(1-|\mu_*(\rho,t)|^2)^{-1}d\rho,$$

where $\mu_*(\ ,t)$ denotes the complex dilatation of $\check f(\ ,t)$.

The differential equation (19.7) has the solution $w = f(s,t)$, where $f(s,1) = f(s)$, $\mu(s,t) = t\mu(s)$, $|s| \le 1$, $0 \le t \le 1$. Hence

$$|f(s) - s| = \left|\int_0^1 f_t(s,t)dt\right| \le \int_0^1 |f_t(s,t)|dt$$

$$< 2\int_0^1 \frac{1}{t}|f(s,t)| \int_{|f(s,t)|}^{1} \frac{(1/r)|\mu_*(r,t)|}{1-|\mu_*(r,t)|^2}drdt.$$

On the other hand, by (19.5) and $\mu(s,t) = t\mu(s)$, $|s| \leq 1$, $0 \leq t \leq 1$, we have

$$(19.8) \quad \mu_*(r,t) = -\mu(\check{f}(r,t), t)\exp\{-2i \arg \check{f}_w(w,t)\}$$
$$= -t\mu(\check{f}(r,t), t)\exp\{-2i \arg \check{f}_w(w,t)\}$$

and, consequently,

$$|f(s) - s| < 2\int_0^1 \int_{|f(s,t)|}^1 \frac{1}{r}|f(s,t)| \frac{|\mu(\check{f}(r,t))|}{1 - t^2|\mu(\check{f}(r,t))|^2} drdt.$$

Since $f = f(\ ,1)$ is a Q-qc mapping, we have $|\mu(\check{f}(r,t))| \leq q = \dfrac{Q-1}{Q+1}$ a.e. in Δ. Hence we obtain

$$|f(s) - s| < 2\int_0^1 \int_{|f(s,t)|}^1 \frac{1}{r}|f(s,t)| \frac{q}{1 - q^2t^2} drdt$$
$$= 2\int_0^1 \frac{q}{1 - q^2t^2}|f(s,t)| \log\frac{1}{|f(s,t)|} dt$$
$$\leq \frac{2}{e}\int_0^1 \frac{q}{1 - q^2t^2} dt = \frac{1}{e}\log\frac{1+q}{1-q} = \frac{1}{e}\log Q,$$

as desired.

Suppose now that there is a constant $c < 1/e$ such that $|f(s) - s| / \log Q \leq c$ for every $f \in E_Q$. Then this estimate holds, in particular, for any function g whose complex dilatation equals $q = (Q-1)/(Q+1)$ in the whole disc Δ. Similarly, $|g(s,t) - s|/\log Q(t) \leq c$ for every function $g(\ ,t) \in E_{Q(t)}$, $0 < t \leq 1$, whose complex dilatation $\varkappa(\ ,t)$ equals tq. Thus we have $Q(t) = (1+qt)/(1-qt)$ and also

$$(19.9) \quad \lim_{t \to 0+} \{|g(s,t) - s|/\log\frac{1+qt}{1-qt}\} \leq c < 1/e,$$

since, as it is easy to show, the limit in (19.9) exists. On the other hand, applying the relation (19.7) with $w = g(s,t)$, we have

$$\lim_{t \to 0+} \frac{f(s,t) - s}{t} = 2\lim_{t \to 0+} \{\frac{f(s,t)}{t} \int_{|f(s,t)|}^1 \frac{(1/r)\varkappa_*(r,t)}{1 - |\varkappa_*(r,t)|^2} dr\},$$

where $\varkappa_*(\ ,t)$ denotes the complex dilatation of $\check{g}(\ ,t)$.

Hence, by (19.8), by the relation $\varkappa(s) = q$, $|s| \leq 1$, and by a well known test concerning the integration of functions depending on a parameter, we get

$$\lim_{t\to 0+}\frac{g(s,t)-s}{t} = -2\lim_{t\to 0+}\{g(s,t)\int_{|g(s,t)|}^{1}\frac{(1/r)\varkappa(\check{g}(r,t))}{1-t^2|\varkappa(\check{g}(r,t))|^2}$$

$$\times\exp(-2i\arg[\check{g}_w(w,t)]_{w=r})dr\} = -2s\int_{|s|}^{1}\frac{q}{r}dr = -2qs\log\frac{1}{|s|}.$$

Consequently, $\lim_{t\to 0+}\dfrac{g(s,t)-s}{qt} = -2s\log\dfrac{1}{|s|}$ and

$$\lim_{t\to 0+}\{[g(s,t)-s]/\log\frac{1+qt}{1-qt}\}$$

$$= \lim_{t\to 0+}\{qt/\log\frac{1+qt}{1-qt}\}\lim_{t\to 0+}\frac{g(s,t)-s}{qt} = -s\log\frac{1}{|s|},$$

which contradicts (19.9) if we set $s=1/e$. Thus our proof is completed.

An analogue of Corollary 13 for Teichmüller mappings reads (Ławrynowicz [6]):

<u>COROLLARY 14</u>. $|f(s)-s|/\log Q<1$ <u>for</u> f <u>being normalized Teichmüller mappings of</u> Δ <u>onto itself with the corresponding</u> Φ <u>as in Theorem</u> 13 <u>and a fixed</u> $t=(Q-1)/(Q+1)$, $1<Q<+\infty$, <u>and</u> $|s|\le 1$. <u>The constant</u> 1 <u>is the best possible</u>.

P r o o f. Let us use the notation of Theorem 13. By this theorem, in particular formula (17.5) with $w=f(s,t)$, we get, as before,

$$(19.10)\quad |f(s)-s| < \frac{1}{\pi}\int_0^t\left|\iint_\Delta \frac{\overline{\Psi_z(z,\tau)}}{\Psi_z(z,\tau)}\frac{dxdy}{f(s,\tau)-z}\right|\frac{d\tau}{1-\tau^2}$$

$$\le \frac{1}{\pi}\int_0^t\iint_\Delta\frac{dxdy}{|f(s,\tau)-z|}\frac{d\tau}{1-\tau^2}.$$

Let $|f(s,\tau)|=r$. It is clear that

$$(19.11)\quad \iint_\Delta |f(s,\tau)-z|^{-1}dxdy = \iint_\Delta |r-z|^{-1}dxdy.$$

But for $0\le r_1\le 1$, $0\le r_2\le 1$ we have

$$\iint_\Delta\frac{dxdy}{|z-r_1|} - \iint_\Delta\frac{dxdy}{|z-r_2|} = \iint_{D_1}\frac{dxdy}{|z-r_1|} - \iint_{D_2}\frac{dxdy}{|z-r_2|}$$

where $D_1 = \Delta \cap \{z: |z - r_1 + r_2| > 1\}$, $D_2 = \Delta \cap \{z: |z - r_2 + r_1| > 1\}$, and, moreover, we have $|z - r_1| < |z + r_2|$ if $z \in \Delta$, $r_1 < r_2$. Hence (19.11) forms a decreasing function of r in $[0; 1]$ and, consequently, it attains its maximum for $r = 0$. Thus

$$(19.12)\qquad \iint_\Delta |r - z|^{-1} dxdy \leq \iint_\Delta |z|^{-1} dxdy = 2\pi .$$

Relations (19.10), (19.11), and (19.12) imply the desired estimate.

In order to prove the second statement we consider for each Q the Teichmüller mapping $g(\ ,t)$, $t = (Q-1)/(Q+1)$, that carries the origin $z = 0$ into a point of the negative real half-axis with the greatest possible distance from the origin within the class in question. This mapping was constructed effectively by Teichmüller [2] and we have (cf. Krzyż [1]):

$$-g(0,t) = \check{\nu}(4\pi^{-1} \operatorname{ar\,coth} Q^{\frac{1}{2}})$$

with ν as in Section 11. On the other hand (cf. e.g. Lehto and Virtanen [1], p. 64, or [2], p. 65), we have $\lim_{r \to 0+} [\nu(r)/\log(4/r)] = 2/\pi$. Therefore

$$\lim_{Q \to 1+} \frac{-g(0,t)}{\log Q} = 4 \lim_{Q \to 1+} \frac{\exp(-2 \operatorname{ar\,coth} Q^{\frac{1}{2}})}{\log Q} .$$

Since $\exp(-2 \operatorname{ar\,coth} Q^{\frac{1}{2}}) = (Q^{\frac{1}{2}} - 1)/(Q^{\frac{1}{2}} + 1)$, then

$$\lim_{Q \to 1+} [-g(0,t)/\log Q] = 1,$$

as desired. (It is worth-while to notice that the mapping $g(\ ,t)$ can easily be deduced from the parametrization method, what we leave as an exercise.)

Next we prove an analogue of Corollary 13 for S_Q (Shah Tao-shing [1]):

COROLLARY 15. *$|f(s) - s|/\log Q < (1/4\pi^2)[\Gamma(\frac{1}{4})]^4 \doteq 4.5$ for $f \in S_Q$, $1 < Q < +\infty$, and $|s| \leq 1$. The constant $(1/4\pi^2)[\Gamma(\frac{1}{4})]^4$ is the best possible.*

Proof. By Theorem 5, in particular formulae (14.1) with $w = f(s,t)$ and (14.2) with $\mu(s,t) = t\mu(s)$, we get, as in the case of Corollary 13,

$$|f(s) - s| < \frac{1}{\pi} \int_0^1 \iint_\Delta \left\{ \frac{|w(1-w)|}{|z(1-z)(w-z)|} + \frac{|w(1-w)|}{|z(1-z)(1-\bar{w}z)|} \right\} dxdy$$

$$\times\, q(1 - q^2 t^2)^{-1} dt \leq M^* \log Q,$$

where $q = (Q-1)/(Q+1)$ and

$$M^* = \frac{1}{2\pi}\max_{w\in\Delta}\iint\left\{\frac{|w(1-w)|}{|z(1-z)(w-z)|} + \frac{|w(1-w)|}{|z(1-z)(1-\bar{w}z)|}\right\}dxdy.$$

We proceed to estimate M^*. Clearly,

$$\iint_{\mathbb{C}\setminus\Delta}\frac{dxdy}{|z(1-z)(w-z)|} = \iint_{\Delta}\frac{dxdy}{|z(1-z)(1-\bar{w}z)|}$$

Hence $M^* = (1/2\pi)\max_{w\in\Delta}\{|w(1-w)|\Phi(w)\}$, where

$$\Phi(w) = \iint_{\mathbb{C}}|z(1-z)(w-z)|^{-1}dxdy.$$

Geometrically, $\Phi(w) = |g(w)[\mathbb{C}]|$, where

$$g(w)(z) = \int[z(1-z)(w-z)]^{-\frac{1}{2}}dz.$$

Therefore $\Phi(w) = 2|\mathrm{im}[P(w)P(1-w)]|$, where

$$(19.13)\qquad P(w) = \int_0^1[t(1-t)(1-wt)]^{-\frac{1}{2}}dt$$

and, consequently, $\Phi(\mathrm{re}\,w) \geq \Phi(w)$ for $w\in\Delta\cap\Delta^1(1)$. Since $|1-1/w| \leq [2(1-\mathrm{re}\,w)]^{\frac{1}{2}}$ for $w\in D = \Delta\cap\{w:\ \mathrm{re}\,w \geq \frac{1}{2}\}$, we obtain

$$\max_{w\in D}\{|1-1/w|\Phi(w)\} = \max_{\frac{1}{2}\leq t\leq 1}\{[2(1-t)]^{\frac{1}{2}}\Phi(t)\} = \Phi(\tfrac{1}{2}).$$

Besides, by $\Phi(1-w) = \Phi(w)$, $w\in\mathbb{C}$, we have

$$\max_{w\in\Delta\cap\Delta^1(1)}\{|w(1-w)|\Phi(w)\} \leq \Phi(\tfrac{1}{2})$$

and, by $\frac{1}{|1-w|}\Phi\left(\frac{1}{|1-w|}\right) = \Phi(w)$, also

$$\max_{w\in\Delta\cap[\mathbb{C}\setminus\mathrm{int}\,\Delta^1(1)]}\{|w(1-w)|\Phi(w)\} = \max_{w\in D}\{|1-1/w|\Phi(w)\} = \Phi(\tfrac{1}{2}).$$

Thus finally we obtain

$$M^* \leq \frac{1}{2\pi}\Phi(\tfrac{1}{2}) = \frac{1}{\pi}\Phi(-1) = \frac{2}{\pi}\left\{\int_0^1\frac{dt}{[t(1-t^2)]^{\frac{1}{2}}}\right\}^2 = \frac{1}{4\pi^2}[\Gamma(\tfrac{1}{4})]^4.$$

The proof that the value obtained is really the best possible may be performed in analogy to that of Corollary 13. We leave it as an exercise.

Now we turn our attention to the problem of estimating $|f(s)|$ in terms of $|s|$ for E_Q, $S_Q^{r,R}$ and S_Q. The result formulated below for S_Q is due to Wang Chuan-fang [1] and improves an earlier result of Mori [1]. The results for $S_Q^{r,R}$ and E_Q are trivial (Ławrynowicz [4]).

<u>COROLLARY 16</u>. $|s|^Q \le |f(s)| \le |s|^{1/Q}$ <u>for</u> $f \in E_Q$ <u>and</u> $|s| \le 1$; <u>hence it follows that a class</u> $S_Q^{r,R}$ <u>is nonempty if</u> $r^Q \le R \le r^{1/Q}$. <u>For</u> $f \in S_Q$ <u>and</u> $|s| \le 1$ <u>we have</u>

$$(19.14) \quad 4^{1-Q}|s|^Q \le |f(s)| \le 4^{1-1/Q}|s|^{1/Q}$$

<u>and - more generally - for</u> $f \in S_Q$ <u>and</u> $|s_1| \le 1$, $|s_2| \le 1$ <u>we have</u>

$$4^{1-Q}\left|\frac{s_1 - s_2}{1 - s_1\bar{s}_2}\right|^Q \le \left|\frac{f(s_1) - f(s_2)}{1 - f(s_1)\overline{f(s_2)}}\right| \le 4^{1-1/Q}\left|\frac{s_1 - s_2}{1 - s_1\bar{s}_2}\right|^{1/Q}.$$

<u>The constants</u> 1, 4^{1-Q}, 1, $4^{1-1/Q}$ <u>and exponents</u> Q, $1/Q$ <u>cannot be improved</u>.

<u>Remark 2</u>. By the geometric characterization of Q-qc mappings in terms of the moduli of annuli it follows that if a class $S_Q^{r,R}$ is nonempty, then $r^Q \le R \le r^{1/Q}$.

<u>Proof of Corollary 16</u>. Applying Lemma 18, in particular condition (v), and setting $\rho = |\mu|$, $\vartheta = \arg\mu$, we get for $|s| \le 1$:

$$\log\frac{1}{|f(s)|} = \int_{|s|}^{1} \operatorname{re}\frac{1+\rho(r)e^{i\vartheta(r)}}{1-\rho(r)e^{i\vartheta(r)}}\frac{dr}{r} = \int_{|s|}^{1}\frac{[1-\rho^2(r)]r^{-1}\,dr}{1+\rho^2(r)-2\rho(r)\cos\vartheta(r)}$$

$$\ge \int_{|s|}^{1}\frac{1-\rho(r)}{1+\rho(r)}\frac{dr}{r} \ge \int_{|s|}^{1}\frac{1}{Q}\frac{dr}{r} = -\frac{1}{Q}\log|s| = \log\frac{1}{|s|^{1/Q}}.$$

Equality holds only for $\mu(r) = -(Q-1)/(Q+1)$, which corresponds to $f(z) = |z|^{1/Q}e^{i\arg z}$, $z \ne 0$, $f(0) = 0$. Applying the result obtained to the inverse function, we have $|f(s)| \ge |s|^Q$. Here the only extremal function for every s is $f(z) = |z|^Q e^{i\arg z}$, $z \ne 0$, $f(0) = 0$, which corresponds to $\mu(r) = (Q-1)/(Q+1)$. The above precise estimates and the relation $E_Q^{r,R}$ $S_Q^{r,R}$ imply that a class $S_Q^{r,R}$ is nonempty if $r^Q \le R \le r^{1/Q}$.

Suppose now that $f \in S_Q$. Clearly,

$$\log\frac{|f(s)|}{|s|^{\tilde{q}(1)}} = -\int_0^1 \frac{\partial}{\partial t}\log|\check{f}(w,t)|^{\tilde{q}(t)}\,dt$$

$$= -\int_0^1 \{\tilde{q}'(t)\log|\check{f}(w,t)| + \tilde{q}(t)\operatorname{re}[\check{f}^{-1}(w,t)\check{f}_t(w,t)]\}dt,$$

where $\tilde{q}$ will be specified later (it has to be continuously differentiable and to satisfy the condition $\tilde{q}(0) = 1$). By Theorem 5, in particular formulae (14.1) with $s = \check{f}(w,t)$ and (14.2) with $\mu(s,t) = t\mu(s)$, we get, as in the case of Corollary 13,

$$\log\frac{|f(s)|}{|s|^{\tilde{q}(1)}} = \int_0^1 \tilde{q}'(t)\log\frac{1}{|\check{f}(w,t)|}dt - \int_0^1 \tilde{q}(t)\operatorname{re}\{\frac{1-\check{f}(w,t)}{\pi}$$

$$\times\iint_\Delta [\frac{\varphi(z,t)}{z(1-z)(\check{f}(w,t)-z)} + \frac{\overline{\varphi(z,t)}}{\bar{z}(1-\bar{z})(1-\check{f}(w,t)\bar{z})}]dxdy\}dt.$$

Since for $z \neq 0$ and $s \neq z, 1/\bar{z}$ we have identically

$$\frac{1-s}{z(1-z)(s-z)} + \frac{1-\bar{s}}{z(1-z)(1-\bar{s}z)} = \frac{1-|s|^2}{z(s-z)(1-\bar{s}z)},$$

then

$$\log\frac{|f(s)|}{|s|^{\tilde{q}(1)}} = \int_0^1 \tilde{q}'(t)\log\frac{1}{|\check{f}(w,t)|}dt$$

$$-\int_0^1 \frac{\tilde{q}(t)}{\pi}\operatorname{re}\iint_\Delta \frac{(1-|\check{f}(w,t)|^2)\,\varphi(z,t)}{z(\check{f}(w,t)-z)(1-\overline{\check{f}(w,t)}z)}dxdydt.$$

Direct calculation (Wang Chuan-fang [1]) gives

$$\iint_\Delta |z(s-z)(1-\bar{s}z)|^{-1}\,dxdy = 4K(|s|)K'(|s|), \quad |s| \leq 1,$$

where $K'(k) = K((1-k^2)^{\frac{1}{2}})$. Therefore, choosing $\mu^*(\ ,t) = t\mu^*$ as the complex dilatation of $\check{f}(\ ,t)$, we obtain

$$\log\frac{|f(s)|}{|s|^{\tilde{q}(1)}} = \int_0^1 \tilde{q}'(t)\log\frac{1}{|\check{f}(w,t)|}dt$$

$$+\frac{2}{\pi}\int_0^1 \frac{q\,\tilde{q}(t)}{1-q^2t^2}(1-|\check{f}(w,t)|^2)\,K(|\check{f}(w,t)|)\,K'(|\check{f}(w,t)|)\,dt$$

$$\leq \log\frac{1}{|s^{**}|}\int_0^1 q'(t)dt + \frac{2}{\pi}(1-|s^*|^2)\,K(|s^*|)\,K'(|s^*|)\int_0^1 \frac{q\,\tilde{q}(t)dt}{1-q^2t^2},$$

where $q = (Q-1)/(Q+1)$, s^* and s^{**} belong to $\check{f}(w, [0;1]]$, and $w = f(s)$.

In order to specify $\tilde{q}$ conveniently we write the obtained estimate in the form

$$(19.15)\quad \log\frac{|f(s)|}{|s|^{\tilde{q}(1)}} \le \int_0^1 [\tilde{q}'(t)\log\frac{1}{|s^{**}|} + \frac{q\tilde{q}(t)}{1-q^2t^2}\log\frac{1}{|s^*|}]dt$$
$$+ [\log|s^*| + \tfrac{2}{\pi}(1-|s^*|^2)\,K(|s^*|)\,K'(|s^*|)]\int_0^1 \frac{q\tilde{q}(t)}{1-q^2t^2}dt.$$

Since (Wang Chuan-fang [1])

$$(19.16)\quad \log k + (2/\pi)(1-k^2)\,K(k)\,K'(k) \le \log 4,\quad 0<k<1,$$

then the second term on the right-hand side of (19.15) is bounded in Δ. On the other hand, if

$$\int_0^1 [\tilde{q}'(t) + 2q(1-q^2t^2)^{-1}\tilde{q}(t)]dt \neq 0$$

then the first term on the right-hand side of (19.15) is unbounded. Thus we choose $\tilde{q}$ so that it satisfies the equation $\tilde{q}'(t) + a(t)\tilde{q}(t) = 0$, $a(t) = 2q(1-q^2t^2)^{-1}$, $0 \le t \le 1$, and by $\tilde{q}(0) = 1$ we get $\tilde{q}(t) = (1-qt)/(1+qt)$. Hence we may take $z^{**} = z^*$. Since $\tilde{q}(1) = 1/Q$, inequalities (19.15) and (19.16) yield the desired estimate of $|f(s)|$ from above and applying this result to the inverse function, we also obtain the desired estimate from below.

The proof that the constants 4^{1-Q}, $4^{1-1/Q}$ and exponents Q, $1/Q$ cannot be improved may be performed in analogy to that of Corollary 13. We leave it as an exercise. The last statement in Corollary 16 is a direct consequence of what we have just proved.

The method of Mori gives in fact a proof of the following, even stronger estimate of $|f(s)|$, $f \in S_Q$, sharp for any $|s|$ and Q:

$$(19.17)\quad \varphi_{1/Q}(|s|) \le |f(s)| \le \varphi_Q(|s|),$$

where $\varphi_Q(r) = \check{\nu}(Q^{-1}\nu(r))$, which was already discovered by Hersch and Pfluger [1]. We have (Ławrynowicz [3]; Hübner [1] in the general case):

$$\varphi_Q(r) \le r\exp[\tfrac{2}{\pi}(1-\tfrac{1}{Q})(1-r^2)K(r)K'(r)] \le 4^{1-1/Q}r^{1/Q},\quad 0<r<1.$$

In turn we quote the following result of Shah Tao-shing and Fan Le-le [1] which improves a result of Lehto, Virtanen and Väisälä [1]:

<u>Suppose that</u> $f\colon D \to D'$ <u>is a</u> Q-<u>qc mapping of a domain</u> D <u>and</u> $\Delta^R(s) \subset D$. <u>Let</u>

$$m_f(s,R) = \min_\alpha |f(s+Re^{i\alpha}) - f(s)|,\quad M_f(s,R) = \max_\alpha |f(s+Re^{i\alpha}) - f(s)|,$$
α *real*.

Then $M_f(s,R) \le \Lambda(R/r, Q)\, m_f(s,r)$ *for* $0<r<R$, *where* $\Lambda(a,Q)$ *is the positive root of the equation*

$$\frac{P(1/\Lambda^2)}{P(1-1/\Lambda^2)} = Q\,\frac{P([(1+a)^{\frac12} - a^{\frac12}]^4)}{P(1-[(1+a)^{\frac12} - a^{\frac12}]^4)},$$ P *being given by* (19.13).

The constant $\Lambda(R/r, Q)$ *cannot be improved*.

For our further purposes in the next section we shall prove this result in a weaker form:

COROLLARY 17. *Suppose that* $f\colon D \longrightarrow D'$ *is a* Q-*qc mapping of a domain* D *and* $\Delta^R(s) \subset D$. *Then* $M_f(s,R) \le \Lambda^*(R/r, Q)\, m_f(s,r)$ *for* $0<r\le R$, *where* $\Lambda^*(a,Q) = 4^{Q-1}\,[(1+a)^{\frac12} - a^{\frac12}]^{-2Q}$.

Proof. Without any loss of generality we may suppose that $f(s) = s = 0$ and $R = M(R) = 1$. Consider a conformal mapping $h\colon \operatorname{int}\Lambda \longrightarrow D \cap \check{f}[\operatorname{int}\Delta]$ satisfying $h(0) = 0$. Since $h[\operatorname{int}\Delta] \supset \Delta$, then, by Koebe's $\frac14$ Theorem, $|h'(0)| \le 4$. Applying now a well known estimate $|h(z)| \le |h'(0)|\,|z|/(1-|z|)^2$ we obtain $|h(z)| \le 4|z|/(1-|z|)^2$ for $|h(z)| \ge r$; hence $|z| \ge [(1+1/r)^{\frac12} - (1/r)^{\frac12}]^2$ for $|h(z)| \ge r$. By Corollary 2 the composed mapping $f\circ h$ can be continued Q-quasiconformally onto Δ and by Corollary 16 it satisfies the estimate $|f\circ h(z_0)| \ge 4^{1-Q} \times [(1+1/r)^{\frac12} - (1/r)^{\frac12}]^{2Q}$, where z_0 is a point at which $|f\circ h|$ attains its minimum on $\operatorname{fr}\Delta^r$, i.e. $|f\circ h(z_0)| = m(r)$. Therefore we arrive at the desired estimate.

In order to prove the general result of Shah Tao-shing and Fan Le-le it is sufficient to utilize in the above consideration the estimate (19.17) instead of (19.14).

Finally we shall prove another result of the same authors [1]:

COROLLARY 18. $M_f(0,R) \le\ (R/r, Q)\, m_f(0,r)$ *for* f S_Q *and* $0<r<R \le 1$, *where*

$$\int_{R/r}^{\lambda(R/r,Q)} \frac{dt}{t\,\eta(t)} = Q,\quad \eta(w) = \frac{1}{2\pi}\iint_{\mathbb{C}} \frac{(1+w)\,dx\,dy}{|z(1-z)(z+w)|} = \frac{1}{\pi}(1+w)P(w)P(1-w),$$

P *being given by* (19.13). *The constant* $\lambda(R/r, Q)$ *cannot be improved*.

Proof. By Theorem 5, in particular formulae (14.1) with $w = f(s,t)$ and (14.2) with $\mu(s,t) = t\,\mu(s)$, we get

$$\frac{\partial}{\partial t}\log\frac{w_2}{w_1} = \frac{1}{\pi}\iint_\Delta \left[\frac{\varphi(z,t)(w_1-w_2)}{z(w_1-z)(w_2-z)} + \frac{\overline{\varphi(z,t)}(w_1-w_2)}{\bar z(1-w_1\bar z)(1-w_2\bar z)}\right]dxdy,$$

where $w_1 = f(re^{i\alpha}, t)$, $w_2 = f(Re^{iA}, t)$, α and A being real. Hence

$$\left|\frac{\partial}{\partial t}\log\left|\frac{w_1}{w_2}\right|\right| = \frac{1}{\pi}\left|\mathrm{re}\iint_\Delta \left[\frac{\varphi(z,t)(w_1-w_2)}{z(w_1-z)(w_2-z)} + \frac{\varphi(z,t)(\bar w_1-\bar w_2)}{z(1-\bar w_1 z)(1-\bar w_2 z)}\right]dxdy\right|$$

$$\le \eta(w_1,w_2)q/(1-q^2t^2),$$

where

$$\eta(w_1,w_2) = \frac{1}{2\pi}|w_1-w_2|\iint_\Delta \left[\frac{1}{|z(w_1-z)(w_2-z)|} + \frac{1}{|z(1-\bar w_1 z)(1-\bar w_2 z)|}\right] \times dxdy$$

and $q = (Q-1)/(Q+1)$. Clearly,

$$\iint_{\mathbb{C}\setminus\Delta} |z(w_1-z)(w_2-z)|^{-1}dxdy = \iint_\Delta |z(1-\bar w_1 z)(1-\bar w_2 z)|^{-1}dxdy.$$

Hence $\eta(w_1,w_2) = \phi(\omega)$, where $\omega = w_1/w_2$ and

$$\phi(\omega) = (1/2\pi)|1-\omega|\iint_{\mathbb{C}} |z(1-z)(\omega-z)|^{-1}dxdy$$

$$= (1/2\pi)|1-R/r|\int_0^{+\infty}\int_{-\pi}^{\pi} |(1-\rho e^{i\vartheta})[(R/r)e^{iA-i\alpha}-\rho e^{i\vartheta}]|^{-1}d\vartheta d\rho.$$

The transformation, given by the formula $s(\rho e^{i\vartheta}) = 1/(1-\rho e^{i\vartheta})$, maps any circle $\partial\Delta^\rho$ onto a circle Γ_ρ. Therefore

$$\phi(\omega) = \frac{1}{2\pi}\int_0^{+\infty}\int_{\Gamma_\rho} |s - 1/(1-\frac{R}{r}e^{iA-i\alpha})|^{-1}|ds|d\rho$$

$$\le \frac{1}{2\pi}\int_0^{+\infty}\int_{\Gamma_\rho} |s-1/(1+\frac{R}{r})|^{-1}|ds|d\rho = \phi(1-\frac{R}{r})$$

and, consequently, $\psi(\omega) = \eta(-\omega)$. But $\eta(w) = \eta(1/w)$, so

$$\left|\frac{\partial}{\partial t}\log\left|\frac{w_2}{w_1}\right|\right| \le \eta\left(\left|\frac{w_2}{w_1}\right|\right)\frac{q}{1-q^2t^2} \quad \text{i.e.} \quad \int_{R/r}^{|w_2/w_1|}\frac{dt}{t\,\eta(t)} \le \log Q,$$

whence the desired estimate follows. The proof that the constant $\lambda(R/r, Q)$ cannot be improved may be performed in analogy to that of Corollary 13. We leave it as an exercise.

<u>In the limit case</u> $R = r$ <u>one obtains the estimate</u> $M_f(0,R) \le \lambda(1,Q) \times m_f(0,r)$ <u>with</u> $\lambda(1,Q) = 1/[\check{\nu}^{-2}(\frac{1}{4}Q^{-1})-1]$ (Lehto, Virtanen and Väisälä [1]).

20. Area distortion under quasiconformal mappings

20. Area distortion under quasiconformal mappings

In this section we are going to prove the following theorem of Gehring and Reich [1]:

THEOREM 17. $\pi^{-1}|f[E]| \le b(Q)(\pi^{-1}|E|)^{a(Q)}$ for $f \in S_Q$, where $a(Q)$ and $b(Q)$ are constants, $a(Q) = Q^{-\hat{a}}$, $1 \le \hat{a} \le 20$, $b(Q) > 0$, and $b(Q) = 1 + \mathcal{O}(Q-1)$ as $Q \to 1+$.

Remark 3. In the paper quoted the constant $\hat{a}$ appearing in the theorem is estimated from above by 40; the estimate by 20 (which is still not the best possible) is due to Reich [2], who had to change the proof considerably (actually the constant appearing in this paper is 17, but this is caused by a mistake, as observed by Kelingos [2]). Finally Gol'dšteĭn [1] has proved that $\hat{a} = 1$. The result without estimating the constants $a(Q)$ and $b(Q)$ is due to Bojarski [1, 2].

Proof of Theorem 17. For greater clarity the proof is divided into nine steps.

Step A. We begin with proving a special case of a result used by Calderón and Zygmund [2] (Lemma 1): If E is a measurable set in $\mathbb{C}$ with $0 < |E| < +\infty$ and $0 < \tau \le 4$, then there is a sequence (D_k) of non-overlapping squares such that $\frac{1}{4}\tau < |E \cap D_k|/|D_k| \le \tau$ for any k, and such that $|E \setminus \bigcup D_k| = 0$.

Clearly, given τ we can find a number $t > 0$ such that

$$(20.1) \qquad (1/t)\int_0^t \chi_E^*(s)\,ds < \frac{1}{4} ,$$

where χ_E^* is an arbitrary nonincreasing function equimeasurable (cf. e.g. Saks [2], p. 143) with the characteristic function χ_E of E. Then for any square D of measure t we have, by (20.1),

$$(20.2) \qquad \frac{1}{|D|}|E \cap D| = \frac{1}{|D|}\iint_D \chi_E(z)\,dxdy \le \frac{1}{t}\int_0^t \chi_E^*(s)\,ds < \frac{1}{4}\tau.$$

Divide now $\mathbb{C}$ into a net of squares of measure t and carry out the following procedure: Divide each square into four equal squares and select those where the average of the function is larger than or equal to $\frac{1}{4}\tau$. Continuing this procedure we obtain a sequence of squares D_k which we claim to have the required properties. Indeed, we obviously have $|E \cap D_k|/|D_k| > - \tau$. Moreover, since any selected square D_k has been obtained by dividing D, where the average of χ_E was less than

$\frac{1}{4}\tau$, we also have $|E\cap D_k| \le |E\cap D| \le \frac{1}{4}\tau|D| = \tau|D_k|$ and therefore $|E\cap D_k|/|D_k| \le \tau$, which together with (20.2) concludes the proof.

Step B. We proceed to prove that if E is as in Step A, $c>1$, and $0<\tau<1$, then there exist a set $U\subset\mathbb{C}$ and functions $g,h\colon \mathbb{C}\to\mathbb{R}_+$ such that

$$(20.3)\qquad |U| \le 2\pi c^2\tau^{-1}|E|,\ g(z)+h(z) \ge |S\chi_E(z)|,\ z\in\mathbb{C},$$

$$(20.4)\qquad \iint_{\mathbb{C}\setminus U} g^2 dxdy \le \tau|E|,\ \iint_{\mathbb{C}\setminus U} h\,dxdy \le K(c)|E|,$$

where the transformation S is considered w.r.t. E and

$$K(c) = 4[(c-1)^{-1} + \log(1-1/c)^{-1}].$$

Let D_k be the squares of Step A and z_k their centres. For any k there is a set E_k such that $E\cap D_k\subset E_k\subset D_k$ and $|E\cap E_k| = \tau|E_k|$. If $D^* = \bigcup D_k$, $E^* = \bigcup E_k$, then

$$|E| = |E\cap D^*| = \Sigma\,|E\cap D_k| > \frac{1}{4}\tau\,\Sigma\,|D_k| = \frac{1}{4}\tau|D^*|$$

and

$$|E| = |E\cap E^*| = \Sigma\,|E\cap E_k| = \tau\,\Sigma\,|E_k| = \tau\,|E^*|.$$

Let

$$(20.5)\qquad g(z) = \tau|S\chi_{E^*}(z)| \text{ and } h(z) = |S\chi_E(z) - \tau S\chi_{E^*}(z)|,\quad z\in\mathbb{C}.$$

By (6.14) we have

$$(20.6)\qquad \iint_{\mathbb{C}} g^2 dxdy = \tau^2|E^*| = \tau|E|.$$

Since $S\chi_E(z) = S\chi_{E^*}(z) = 0$ for a.e. $z\,\bar{\in}\,E^*$, then

$$h(s) = (1/\pi)\Big|\sum_{k=1}^{+\infty}\iint_{E_k}[\chi_E(z) - \tau\chi_{E^*}(z)](z-s)^{-2}\,dxdy\Big|.$$

On the other hand

$$\iint_{E_k}(\chi_E - \tau\chi_{E^*})dxdy = |E\cap E_k| - \tau|E_k| = 0,\ k=1,2,\dots$$

Therefore

$$\iint_{E_k}[\chi_E(z) - \tau\chi_{E^*}(z)]\frac{dxdy}{(z-s)^2} = \iint_{E_k}[\chi_E(z) - \tau\chi_{E^*}(z)] \times\Big[\frac{1}{(z-s)^2} - \frac{1}{(z_k-s)^2}\Big]dxdy.$$

Now, let $U_k = \Delta^{cr_k}(z_k)$, $r_k = \frac{1}{2}\operatorname{dia} D_k$, and $U = \bigcup U_k$. If $z\in D_k$, $s\,\bar{\in}\,U_k$, we have

$$\left|\frac{1}{(z-s)^2}-\frac{1}{(z_k-s)^2}\right| = \left|\frac{z-z_k}{(z_k-s)^3}\right|\left|\frac{z_k-s}{z-s}\right|\left(1+\left|\frac{z_k-s}{z-s}\right|\right)$$

$$\leq \frac{r_k}{|s-z_k|^3}\frac{|s-z_k|}{|s-z_k|-r_k}\left(1+\frac{|s-z_k|}{|s-z_k|-r_k}\right).$$

Thus

$$h(s) \leq (1/\pi)\, M_k(s) \iint_{E_k} (\chi_E - \tau\chi_{E*})\,dxdy$$

where

$$M_k(s) = \sum_{k=1}^{+\infty} \frac{r_k}{|s-z_k|^3}\frac{|s-z_k|}{|s-z_k|-r_k}\left(1+\frac{|s-z_k|}{|s-z_k|-r_k}\right).$$

On the other hand

$$\iint_{\mathbb{C}\setminus U} M_k(z)\,dxdy \leq \iint_{\mathbb{C}\setminus U_k} M_k(z)\,dxdy$$

$$= r_k \int_{cr_k}^{+\infty}\int_0^{2\pi} \frac{\rho}{\rho - r_k}\left(1+\frac{\rho}{\rho-r_k}\right)d\vartheta\,\frac{d\rho}{\rho} = \tfrac{1}{2}\pi\, K(c).$$

Also,

$$\iint_{E_k} |\chi_E - \tau\chi_{E*}|\,dxdy = \iint_{E_k\setminus E} |0-\tau|\,dxdy + \iint_{E_k\cap E} |1-\tau|\,dxdy$$

$$= \tau|E_k\setminus E| + (1-\tau)|E_k\cap E| = \tau(|E_k| - |E_k\cap E|) + (1-\tau)|E_k\cap E|$$

$$= 2\tau(1-\tau)|E_k| < 2\tau|E_k|.$$

Therefore

$$\iint_{\mathbb{C}\setminus U} h\,dxdy \leq K(c)\tau\sum_{k=1}^{+\infty}|E_k| = K(c)\tau|E^*| = K(c)|E|,$$

what, together with (20.6), yields (20.4). Finally, relations (20.3) follow from (20.5) and

$$|U| \leq \Sigma|U_k| = \pi c^2\,\Sigma r_k^2 = \tfrac{1}{2}\pi c^2\,\Sigma|D_k| < 2\pi c^2\tau^{-1}\,\Sigma|E\cap D_k|$$
$$= 2\pi c^2\tau^{-1}|E|.$$

Step $\underline{C}$. Consider now the distribution function m_E defined by $m_E(t) = |\{z: |S\chi_E(z)| \geq t\}|$, $t>0$. From (6.14) we have

(20.7) $\quad m_E(t)/|E| \leq 1/t^2.$

<u>We claim that</u>

(20.8) $m_E(t)/|E| < 40/t.$

To this end we choose U, g and h as in Step B. Let

$$G(c,t,\tau) = |\{z\colon g(z) \geq t,\ z \in U\}|,\quad H(c,t,\tau) = |\{z\colon h(z) \geq t,\ z \in U\}|.$$

By (20.4) we find that $G(c,t,\tau) \leq t^{-2}|E|$, $H(c,t,\tau) \leq K(c)t^{-1}|E|$. Hence, for any $a > 0$ we have

$$\text{(20.9)}\quad \begin{aligned} m_E(t) &\leq |\{z\colon g(z) + h(z) \geq t\}| \leq |U| + |\{z\colon g(z) + h(z) \geq t,\ z \in U\}| \\ &\leq |U| + G(c, at, \tau) + H(c, t - at, \tau) \\ &\leq [2\pi c^2\tau^{-1} + a^{-2}\tau t^{-2} + (1-a)^{-1} K(c)t^{-1}]|E|. \end{aligned}$$

Choosing $\tau = (2\pi)^{\frac{1}{2}}act$ for $0 < t \leq 4/(2\pi)^{\frac{1}{2}}ac$, we get

$$m_E(t)/|E| \leq [2^{\frac{3}{4}}\pi^{\frac{1}{2}}a^{-1}c + (1-a)^{-1}K(c)]/t.$$

To minimize the right-hand side we choose $a = 1/[1 + 2^{-\frac{3}{4}}\pi^{-\frac{1}{4}}c^{-\frac{1}{2}}K^{\frac{1}{2}}(c)]$. Hence

$$m_E(t)/|E| \leq [2^{-}\pi^{\frac{1}{2}}c^{\frac{1}{2}} + K^{\frac{1}{2}}(c)]^2/t,$$

what for $c = 2\frac{2}{3}$ yields (20.8) provided that $0 < t < 1\frac{9}{100}$. Finally, applying (20.7), which is valid for all $t > 0$, we conclude (20.4) also for all $t > 0$.

S t e p <u>D</u>. <u>If</u> $E = \Delta$ (or, more generally, E is any disc), <u>we have</u> $S\chi_E(z) = (1/\pi z^2)[1 - \chi_E(z)]|E|$ <u>and</u> $m_E(t)/|E| = 1/t$ <u>for</u> $0 < t \leq 1$, $m_E(t) = 0$ <u>for</u> $t > 1$, <u>whence</u>

$$\sup_t \sup_E \{t\, m_E(t)/|E|\} \geq 1$$

(Reich [2] has shown more: $\liminf_{t \to 0+} \sup_E \{t\, m_E(t)/|E|\} > 1$).

S t e p <u>E</u>. Next we are going to prove that, <u>for any measurable set</u> $E \subset \mathbb{C}$ <u>and</u> $\alpha \in (0; 1]$, <u>we have</u>

$$\text{(20.10)}\quad \int_\alpha^1 m_E\, dt \leq 20(1 + \log \alpha^{-1})|E|.$$

The estimate (20.9) yields

$$\text{(20.11)}\quad \int_0^t y\, m_E(y)\, dy \leq \pi c^2\tau^{-1}t^2|E|$$

$$+ a^{-2} \int_0^{at} y\, G(c,y,\tau)dy + (1-a)^{-2} \int_0^{(1-a)t} y\, H(c,y,\tau)dy.$$

We proceed to estimate the last two integrals. By (20.5) and (20.4) we have

$$\int_0^{at} y\, G(c,y,\tau)dy \leq \int_0^{+\infty} y\, G(c,y,\tau)dy = \int_0^{+\infty} G(c,y^{\frac{1}{2}},\tau)dy = \iint_{\mathbb{C}\setminus U} g^2 dxdy \leq \tau|E|.$$

To obtain an upper bound for the second integral on the right-hand side of (20.11) we utilize the fact that $H(c, \ ,\tau)$ is a decreasing function and that, by (20.4),

$$\int_0^{(1-a)t} H(c,y,\tau)dy \leq \int_0^{+\infty} H(c,y,\tau)dy = \iint_{\mathbb{C}\setminus U} h\, dxdy \leq K(c)|E|.$$

Subject to the above restrictions the largest possible value for the integral in question is obtained for $H(c,y,t) = (1-a)^{-1} K(c)t^{-1}|E|$, whence

$$\int_0^{(1-a)t} y\, H(c,y,\tau)dy \leq (1-a)^{-1} K(c)t^{-1}|E| \int_0^{(1-a)t} y\, dy$$
$$= \tfrac{1}{2}(1-a)\, K(c)\, t\, |E|.$$

Consequently, from (20.11) we obtain

$$\int_0^t y\, m_E(y)\, dy \leq [\pi c^2 \tau^{-1} t^2 + a^{-2}\tau + (1-a)^{-1} K(c)\, t]|E|. \tag{20.12}$$

By Step B (20.12) is valid as long as $0<\tau\leq 1$, $a\geq 0$, and $c>1$. The right-hand side of (20.12) is just t^2 times the right-hand side of (20.9). Hence, with the same choice of τ, a, and c as in Step C, we conclude, analogously, that

$$2t^{-2} \int_0^t y\, m_E(y)\, dy / |E| < 40/t, \quad 0\leq t\leq 1. \tag{20.13}$$

Therefore we get for $0<\alpha\leq 1$

$$\int_\alpha^1 t^{-2} \int_0^t y\, m_E(y)\, dydt = \int_0^\alpha \int_\alpha^1 dtdy + \int_\alpha^1 \int_y^1 dtdy < 20|E|\log \alpha^{-1}$$

and

$$\int_\alpha^1 m_E(y)\, dy < 20|E|\log\alpha^{-1} + \int_\alpha^1 y\, m_E(y)dy + \int_0^\alpha y\, m_E(y)dy - \alpha^{-1}\int_0^\alpha y\, m_E(y)dy$$

$$\leq 20|E|\log{}^{-1} + \int_0^1 y\, m_E(y)\, dy.$$

Applying again (20.11), namely: to the last term, we arrive at the desired estimate.

S t e p F. In turn we claim that <u>there is a positive constant</u> $\beta \leq 19$ <u>such that if</u> $|E| \leq \pi$ <u>then</u>

(20.14) $\quad \|S\chi_E\|_1 \leq 20|E|\log(\pi/|E|) + \beta|E|,$

<u>the norm</u> $\|\ \|_1$ <u>being taken w.r.t.</u> Δ. Let $\mu_E(t) = |\{z: |S\chi_E(z)| \geq t,\ z \in \Delta\}|$. Clearly, $\mu_E(t) \leq \min[\pi, m_E(t)]$ and

$$\|S\chi_E\|_1 = \int_0^{+\infty} \mu_E(t)\,dt.$$

Hence, by (20.7) and Step E, we obtain for $0<\alpha<1$

$$\|S\chi_E\|_1 \leq \int_0^{\alpha} \pi\,dt + 20|E|\log\alpha^{-1} + 20|E| + \int_1^{+\infty} |E|t^{-2}\,dt.$$

The particular choice $\alpha = (1/\pi)|E|$ gives (20.14) with $\beta = 19$.

S t e p G. <u>We proceed then to apply Theorem</u> 5, <u>in particular formulae</u> (14.1) <u>with</u> $w = f(s,t)$ <u>and</u> (14.2) <u>with</u> $\mu(s,t) = t\,\mu(s)$, <u>where</u> μ <u>is the complex dilatation of</u> f. By this theorem and Lemma 4, for every t, $0 \leq t \leq 1$, the differential equation (14.1) can be expressed in the form

(20.15) $\quad w_t = \frac{1}{\pi}\iint_\Delta \varphi(z,t)\left(\frac{1}{w-z} + \frac{1}{z}\right)dxdy + h^*(w,t),$

where $h^*(\ ,t)$ is holomorphic in $\operatorname{int}\Delta$. Let $G(w,t)$ denote the right-hand side of (20.15). By Theorem 7 there is a constant $\beta^* > 0$ such that

(20.16) $\quad |\operatorname{re} h_w^*(w,t)| \leq \frac{1}{\pi}\left|\iint_\Delta G_{\bar z}(z,t)\frac{w^2dxdy}{\bar z(1-w\bar z)}\right| \leq \beta^*\frac{q}{1-q^2t^2}$

whenever $|w| \leq \frac{1}{2}$ and $0 \leq t \leq 1$, where $q = (Q-1)/(Q+1)$.

Let now E be a measurable set in Δ. Let $f[E,t) = \{w: w = f(z,t),\ z \in E\}$. Since

$$\frac{d}{dt}|f[E,t)| = \frac{d}{dt}\iint\{|f_z(z,t)|^2 - |f_{\bar z}(z,t)|^2\}dxdy$$
$$= 2\iint_{f[E,t)} \operatorname{re} G_z(z,t)\,dxdy,$$

then, by (20.15),

$$\frac{d}{dt}|f[E,t)| = 2 \iint_{f[E,t)} \mathrm{re}[S\, G_{\bar z}(\ ,t)(z) + h^*_{\bar z}(z,t)]dxdy,$$

where the transformation S is considered w.r.t. Λ. But for any bounded function $h: f[E,t) \longrightarrow \mathbb{C}$ with compact support we have

$$\iint_{f[E,t)} S\, G_{\bar z}(\ ,t)(z)\, h(z)\, dxdy = \iint_{f[E,t)} G_{\bar z}(z,t)\, Sh(z)\, dxdy.$$

Hence, for $h = \chi_{f[E,t)}$, we obtain, by (14.2) with $\mu(s,t) = t\,\mu(s)$,

$$(20.17)\quad \frac{d}{dt}|f[E,t)| \le 2\iint |S\chi_{f[E,t)}(z)| \frac{q}{1-q^2t^2} dxdy + 2\iint_{f[E,t)} \mathrm{re}\, h^*_{\bar z}(z,t)\, dxdy.$$

S t e p H. For further rearrangement of the estimate (20.17) we apply Corollary 16, especially the second inequality in (19.14). Suppose first that E is a measurable set in $\{z: |z| \le 8^{-Q}\}$. Then, since $f \in S_Q$, $f[E,t)$ lies in $\Lambda^{\frac{1}{2}}$. From Steps G and F, especially formulae (20.16), (20.17), and (20.14), we have

$$\frac{d}{dt}|f[E,t)| \le |f[E,t)|\,[20 \log \frac{\pi}{|f[E,t)|} + (\beta + \beta^*)] \frac{2q}{1-q^2t^2}$$

for $0 \le t \le 1$, and with a change of variables and integration we get

$$\pi^{-1}|f[E,t)| \le \exp\{\tfrac{1}{20}(\beta + \beta^*)[1 - (\tfrac{1-qt}{1+qt})^{20}]\}\{\pi^{-1}|f[E,0)|\}^{(\frac{1-qt}{1+qt})^{20}}$$

Setting $t = 1$ we obtain the statement of Theorem 17 with $a(Q) = Q^{-\hat a}$, $a = 20$, $b(Q) > 0$, and $b(Q) = b_0(Q)$, where

$$(20.18)\quad b_0(Q) = \exp[\tfrac{1}{20}(\beta + \beta^*)(1 - Q^{-20})] = 1 + (\beta + \beta^*)(Q-1) + o(Q-1)$$

as $Q \longrightarrow 1+$.

S t e p I. Finally we are going to rearrange correspondingly the estimate (20.17) for any measurable set E in Λ on applying Corollary 17. To this end we extend the mapping f in question to f^* S^*_Q by the formulae $f^*(s) = f(s)$ for $|s| \le 1$ and by $f^*(s) = 1/\overline{f(1/\bar s)}$ for $1 < |s| < +\infty$, $f^*(\infty) = \infty$. Then, by Corollary 17,

$$(20.19)\quad M_{f^*}(0,r) \le \Lambda^*(1,Q)\, m_{f^*}(0,r) \quad \text{for } 0 < r < +\infty,$$

in particular for $r > 1$. Clearly,

(20.20) $m_{f^*}(0,r) \le r^Q, \quad M_{f^*}(0,r) \ge r^{1/Q}$,

(20.21) $\Lambda^*(1,Q) = 1 + \mathcal{O}(Q-1)$ as $Q \to 1+$.

Consider, in analogy to the proof of Corollary 17, a conformal mapping $h\colon \operatorname{int}\Delta \to \operatorname{int}\Delta^r$ satisfying $h(0) = 0$. The Schwarz lemma applied to $z \mapsto \check{h}(m_{f^*}(0,r)\,z)$ and to $z \mapsto M_{f^*}^{-1}(0,r)h(z)$ gives for $z \in \operatorname{int}\Delta$

(20.22) $|h(z)| \ge m_{f^*}(0,r)\,|z|$

and, for $z \in \check{h}(\operatorname{int}\Delta)$, by (20.22),

(20.23) $$|h'(z)| \le M_{f^*}(0,r)\frac{|M_{f^*}^{-1}(0,r)h(z)|^2}{1-|z|^2} \le M_{f^*}(0,r) + 2\frac{M_{f^*}(0,r) - m_{f^*}(0,r)}{m_{f^*}^2(0,r) - 1}.$$

Let now $r = \{\max[8,\ 3^{\frac{1}{2}}\Lambda^*(1,Q)]\}^Q$. Then (20.19) and (20.20) imply that $m_{f^*}(0,r) \ge 3^{\frac{1}{2}}$, and with (20.23) we conclude that

(20.24) $$|f[E]| = \iint\limits_{\check{h}\circ f[E]} |h'(z)|^2 dxdy \le M_{f^*}^2(0,r)\frac{M_{f^*}^2(0,r)}{m_{f^*}^2(0,r)}|\check{h}\circ f[E]|.$$

If we apply the statement of Step H to the Q-qc mapping $s \mapsto \check{h}\circ f^*(rs)$, we obtain

(20.25) $\pi^{-1}|\check{h}\circ f[E]| \le b_0(Q)(\pi^{-1}r^{-2}|E|)^{Q^{-20}}$,

where $b_0(Q)$ is given by (20.18). Finally, combining (20.19), (20.20), (20.24), and (20.25), we obtain the statement of Theorem 17 with

(20.26) $b(Q) = b_0(Q)[\Lambda^*(1,Q)]^4\, r^{2(Q-Q^{-20})}$.

Since, by (20.18) and (20.21), each factor on the right-hand side of (20.26) is of the form $1+\mathcal{O}(Q-1)$ as $Q \to 1+$, we conclude that $b(Q)$ is also of this form. This completes the whole proof.

A similar problem of estimating distortion of hyperbolic area (a notion analogous to the hyperbolic distance) under qc mappings has been investigated by Gehring [4], Kelingos [2], and Gol'dšteĭn [1].

21. Parametrical methods for conformal mappings with quasiconformal extensions

Qc homeomorphisms of the plane which are conformal in some part of it, play an important role in the theory of Teichmüller spaces. On the other hand such homeomorphisms have a close connection with the classical theory of univalent functions. In particular a series of quite natural problems arises while reformulating classical problems concerning univalent functions for the above mentioned class of homeomorphisms. Therefore it is quite understandable that a considerable interest has recently arisen in investigating univalent functions with qc extension.

Bojarski's integral formula for homeomorphic solutions of the Beltrami equation (also called I. N. Vekua's integral formula) with the complex dilatation of compact support provided an example for mappings of this kind. However the paper of Ahlfors and Weill [1] was perhaps mainly responsible for arising the interest in such homeomorphisms. Another sufficient conditions for univalent functions to have a qc extension were also obtained by Ahlfors [8], Becker [1-6], Becker and Pommerenke [1], Duren and Lehto [1], and others.

Pioneering research in this direction has been done by Kühnau who in a series of papers [1-13] (cf. also Kruškal [6], Kühnau and Dittmar [1], Kühnau and Niske [1], Kühnau and Thüring [1], Kühnau and Blaar [1], Kühnau and Hoy [1], Kühnau and Timmel [1], Kühnau and Baumgarten [1], and Schober [2]) has given many interesting results suggested by known theorems in the theory of conformal mapping and valid for univalent functions admitting a qc extension. The latter mappings appear in Kühnau's papers as a particular case of mappings with bounded local dilatation (i.e. dilatation at a point), the bound being a function given in advance. Some interesting results in this direction have been also obtained by Lehto [3, 6], Lehto and Tammi [1], Schiffer and Schober [1, 2, 4, 5], Krzyż [3, 4], Fait, Krzyż and Zygmunt [1], Krzyż and Soni [1], and others.

In previous sections we were concerned with quasiconformal mappings generated by the complex dilatation $\mu(\,,t)$ depending on a real parameter t. Another approach is due to Lehto who considered families of qc mappings depending on a complex parameter w with dilatation $\mu(\,,w)$ being a holomorphic function of the parameter w; cf. Lehto [4-8] (complex t appeared already in Ahlfors' book [5]).

If $f(\,,w)$ is a suitably normalized qc selfmapping of the plane whose complex dilatation $\mu(\,,w)$ depends holomorphically on w, then the mapping $f(\,,w)$ itself also depends holomorphically on w for almost all z. This leads to a general majorization principle also due to Lehto, which has many interesting applications; cf. Lehto [4-6] and also Göktürk [1]. It is easily seen that a suitable normalization of $f(\,,w)$ is necessary. E.g. $f(z,w) = (z + w\bar{z})/(1 - w\bar{w})$, $|w| \leq q < 1$, is a qc selfmapping of $\mathbb{E}$ whose complex dilatation $\mu(\ ,w)$ is holomorphic, however $f(z,\ )$ is not holomorphic for $z \neq 0$.

Suppose that μ is a complex-valued function defined on $\mathbb{C} \times A$, where A is a domain in the finite plane $\mathbb{C}$. Suppose, moreover, that for any fixed $w \in A$, $\mu(\ ,w)$ is measurable, $\operatorname{spt}\mu(\ ,w) \subset A$ and $\|\mu(\ ,w)\|_\infty < 1$. *Let $\Sigma(Q)$, $1 \le Q < +\infty$, be the class of qc mappings of $\mathbb{E}$ onto itself whose restriction to $\Delta^* = \{z: |z| > 1\}$ is meromorphic with a simple pole at ∞ and the expansion*

$$(21.1)\qquad f(z) = z + \sum_{n=1}^{+\infty} b_n z^{-n}, \quad |z| > 1,$$

(i.e. $f|\Delta^*$ is an element of the familiar class Σ), *while the complex dilatation $\mu = f_{\bar z}/f_z$ satisfies the condition* (7.3). From Theorem 2 it follows that each $\mu(\ ,w)$ with the above stated properties generates for each $w \in A$ a unique function $f(\ ,w) \in \Sigma(Q)$, where Q is given by (9.2).

Lehto's method is based on the following (Lehto [4, 7]):

THEOREM 18. *Suppose that $z \mapsto \mu(z,w)$ satisfies the above stated conditions and is holomorphic in A for every fixed $z \in \mathbb{C}$. If $f(\ ,w)$ is a function of $\Sigma(Q)$ generated by $\mu(\ ,w)$, then $w \mapsto f(z,w)$ is holomorphic in A for every finite z.*

P r o o f. Let F be a compact subset of A. We shall prove that

$$(21.2)\qquad \sup_{w \in F} \|\mu(\ ,w)\|_\infty < 1.$$

Assume that this is not true. Then there is a sequence of $w_n \in F$ such that $w_n \to a \in F$, whereas $\|\mu(\ ,w_n)\|_\infty \to 1-$. Obviously,

$$(21.3)\qquad \|\mu(\ ,w_n)\|_\infty \le \|\mu(\ ,w_n) - \mu(\ ,a)\|_\infty + \|\mu(\ ,a)\|_\infty .$$

Since the functions $\mu(z,\ )$ are holomorphic and uniformly bounded, they form a family of equicontinuous functions; cf. e.g. Ahlfors [6], p. 216. Hence, the first term on the right in (21.3) tends to 0 as $n \to \infty$ and, consequently, $\|\mu(\ ,a)\|_\infty = 1$ in contradiction with our assumptions on μ.

Our further arguments are based on Bojarski's method for the solution of Beltrami's equation (cf. Chapter I).

Let F be a compact subset of A and let

$$q = \sup_{w \in F} \|\mu(\ ,w)\|_\infty = q(F) < 1.$$

21. Conformal mappings with quasiconformal extensions

We choose $p>2$ so that the norm of the Hilbert transformation $\|S\|_p$ in L^p satisfies $q\|S\|_p<1$. In what follows we shall apply the T-transformation defined in (4.5) to functions of two (or more) variables, however, we shall be concerned with the operator T acting on the first variable only so that e.g.

$$T\,g(z,w) = -(1/\pi)\iint_{\mathbb{C}} g(\zeta,w)(\zeta-z)^{-1}\,d\xi\,d\eta\,.$$

Similar notation is adapted for the Hilbert transformation:

$$S\,g(z,w) = -(1/\pi)\iint_{\mathbb{C}} g(\zeta,w)(\zeta-z)^{-2}\,d\xi\,d\eta\,.$$

With this notation the function $f(\;,w)\in\Sigma(Q)$, generated by $\mu(\;,w)$, $w\in F$, has, according to our previous considerations, the following form:

$$(21.4)\qquad f(z,w) = z + \sum_{n=1}^{+\infty} T\varphi_n(z,w),$$

where

$$(21.5)\qquad \varphi_1(z,w) = \mu(z,w),\quad \varphi_n(z,w) = \mu(z,w)\,S\,\varphi_{n-1}(z,w),\quad n=2,3,\ldots$$

Since for ϕ of the class L^p, $\operatorname{spt}\phi\subset\Delta$, we have

$$(21.6)\qquad |T\phi(z)| \le C\|\phi\|_p,$$

where C depends on p only, and since $\Sigma\varphi_n$ is uniformly absolutely convergent in L^p for $w\in F$, we see that the series (21.4) is uniformly convergent w.r.t. $w\in F$ for any fixed z. Thus $f(z,\;)$ is holomorphic in A if we prove that each $T\varphi_n(z,\;)$ is holomorphic there.

To this end we define a sequence of functions $\phi_n(z,w)$ that arises from (φ_n) by formal differentiation w.r.t. w:

$$(21.7)\qquad \begin{aligned} &\phi_1(z,w) = \mu'(z,w),\\ &\phi_n(z,w) = \mu(z,w)\,S\phi_{n-1}(z,w) + \mu'(z,w)\,S\,\varphi_{n-1}(z,w),\quad n=2,3,\ldots, \end{aligned}$$

where the prime indicates differentiation w.r.t. w. Let now F be a closed disc contained in A with centre w and let h be so chosen that $w+h\in F$. We choose $p>2$ such as before, i.e. $q(F)\|S\|_p<1$, and set

$$(21.8)\qquad \Lambda_n(z,w,h) = \frac{1}{h}\,[T\varphi_n(z,w+h) - T\varphi_n(z,w)] - T\phi_n(z,w).$$

The holomorphy of $T\varphi_n(z,\;)$ is equivalent to the relation

$$\lim_{h\to 0} \Delta_n(z,w,h) = 0$$

for any fixed positive integer n, where $z \in \mathbb{C}$. From (21.8) it is evident that

(21.9) $\Delta_n(z,w,h) = T\rho_n(z,w,h)$,

where

(21.10) $\rho_n(z,w,h) = \frac{1}{h}[\varphi_n(z, w+h) - \varphi_n(z,w)] - \psi_n(z,w)$.

Thus all we need to prove is

(21.11) $\lim_{h\to 0} \|\rho_n(\;,w,h)\|_p = 0$

which clearly implies $\lim_{h\to 0} \Delta_n(z,w,h)$ by (21.6), (21.9), and (21.10).

Now

$$\rho_n(z,w,h) = \frac{1}{h}[\varphi_n(z, w+h) - \varphi_n(z,w)] - \mu(z,w)\, S\psi_{n-1}(z,w)$$
$$- \mu'(z,w)\, S\varphi_{n-1}(z,w).$$

Using the expression (21.5) for φ_n we see that

(21.12) $\rho_n(z,w,h) = \alpha_n(z,w,h) + \beta_n(z,w,h) + \gamma_n(z,w,h)$,

where

(21.13) $\alpha_n(z,w,h) = \mu(z,w)\, S\rho_{n-1}(z,w,h)$,

(21.14) $\beta_n(z,w,h) = \{\frac{1}{h}[\mu(z, w+h) - \mu(z,w)] - \mu'(z,w)\}\, S\varphi_{n-1}(z, w+h)$,

(21.15) $\gamma_n(z,w,h) = \mu'(z,w)\, S[\varphi_{n-1}(z, w+h) - \varphi_{n-1}(z,w)]$.

Since $(\varphi_n(z, \;))$ is uniformly bounded on F, so is $(S\varphi_n(z, \;))$. Application of the Cauchy integral formula to

$$\rho_1(z,w,h) = \frac{1}{h}[\mu(z, w+h) - \mu(z,w)] - \mu'(z,w)$$

shows that

$$\lim_{h\to 0} \|\rho_1(\;,w,h)\|_\infty = 0.$$

Therefore, by (21.14)

$$\lim_{h\to 0} \|\beta_n(\;,w,h)\|_\infty = 0 \text{ for fixed } w \in F,\ n = 1,2,\ldots,$$

and also, by (21.15),

$$\lim_{h\to 0} \|\gamma_2(\ ,w,h)\|_p = 0.$$

Hence, by (21.12)-(21.15), $\lim_{h\to 0} \|\ _2(\rho, w, h)\|_p = 0$ and, consequently, by (21.10),

$$\|\varphi_2(\ , w+h) - \varphi_2(\ ,w)\|_p \to 0$$

so that

$$\|\alpha_3\|_p \to 0 \quad \text{and} \quad \|\gamma_3\|_p \to 0, \quad \text{i.e.} \quad \|\rho_3\|_p \to 0.$$

An obvious induction yields

$$\lim_{h\to 0} \|\rho_n(\ ,w,h)\|_p = 0$$

uniformly on F for any fixed positive integer n and thus proves Theorem 18.

In some particular cases the holomorphy of $f(z,\ )$ can be verified in a straightforward manner. Let μ be a measurable function in $\mathbb{C}$ satisfying (7.3) and $\operatorname{spt}\mu \subset \Lambda$. Consider the function

$$f(z) = z + \sum_{n=1}^{+\infty} b_n z^{-n}, \quad f \in \Sigma(Q),$$

generated by μ and also the function $z \mapsto f(z,w)$ generated by $\mu(z,w) = w\mu(z)$, $|w| < (Q-1)/(Q+1)$. By (7.3), for any $w \in \operatorname{int}\Lambda$, we have $f(\ ,w) \in \Sigma(Q)$. From (21.5) it follows that $\varphi_n(z,w) = w^n \varphi_n(z)$ and, consequently, by (21.4):

$$f(z,w) = z + \sum_{n=1}^{+\infty} w^n T\varphi_n(z) = z + \sum_{n=1}^{+\infty} b_n w^n z^{-n}.$$

The power series in w on the right is convergent at $w = 1$ for any $z \in \Lambda^*$ so that $f(z,\ )$ is holomorphic in $\operatorname{int}\Lambda$.

<u>COROLLARY 19</u>. <u>Suppose that</u>

$$f(z,w) = z + \sum_{n=1}^{+\infty} b_n(w) z^{-n}, \quad |z| > 1. \tag{21.16}$$

<u>If</u> $\mu(z,\ )$ <u>is holomorphic in</u> A <u>for any finite</u> z, <u>then the functions</u> b_n, $n = 1,2,\ldots$, <u>are holomorphic</u> (<u>in</u> A).

<u>Proof</u>. We have

$$b_1(w) = \lim_{z\to\infty} z[f(z,w) - z]. \tag{21.17}$$

By means of the area theorem: $\sum_{n=1}^{+\infty} n|b_n(w)|^2 \le 1$ and the Schwarz inequality we can prove that

$$|z[f(z,w) - z]| = |\sum_{n=2}^{+\infty} b_n(w)z^{-n+1}| = |\sum_{n=2}^{+\infty} n^{\frac{1}{2}} b_n(w)\, n^{-\frac{1}{2}} z^{-n+1}|$$

$$\le 1/(|z| - 1),$$

so that the convergence in (21.17) is uniform in w as $z \to \infty$. Since $z[f(z,\ ) - z]$ is holomorphic according to Theorem 18, so is b_1. A similar procedure applied to $z^2[f(z,\ ) - z - z^{-1}b_1]$ shows that b_n is holomorphic (in A). An obvious induction yields holomorphy of b_n for all positive integers n.

COROLLARY 20. The functions $w \mapsto f_z^{(n)}(z,w)$, $z \subset \Delta^*$, $n = 1,2,\dots$, are holomorphic in A.

Proof. According to (21.16) $f(\ ,w)$ is the sum of a Laurent series in Δ^*, so we can differentiate both sides n times with respect to z. After differentiation $f_z^{(n)}(z,\ )$ is represented as a sum of a series of holomorphic functions of w, almost uniformly convergent in A which means that $f_z^{(n)}(z,\ )$ is holomorphic in A.

Let $S(Q)$ be the class of quasiconformal selfmappings f of the extended plane $\mathbb{E}$ whose restrictions to $\operatorname{int}\Delta$ are holomorphic and univalent functions of the familiar class S (i.e. $f(z) = z + a_2z^2 + \dots$ for $|z| < 1$), whereas the complex dilatations μ of $f \in S(Q)$ satisfy (7.3). Given a measurable function μ with $\operatorname{spt}\mu \subset \operatorname{cl}\Delta^*$ and (7.3) fulfilled, a unique $f \in S(Q)$ can be found only if a suplementary condition is added, e.g. $f(\infty) = \infty$. In fact, by Theorem 2 there exists a unique f_0 with given complex dilatation μ_0 leaving $0, 1, \infty$ invariant. Since $\operatorname{spt}\mu_0 \subset \Delta^*$ and f_0 is holomorphic and univalent in $\operatorname{int}\Delta$, $f(z) = f_0(z)/f_0'(0)$ is in $S(Q)$ and satisfies $f(\infty) = \infty$. The subclass of all $f \in S(Q)$ that satisfy $f(\infty) = \infty$ is denoted by $S^\infty(Q)$.

Theorem 18 has a counterpart for the class $S^\infty(Q)$. Let $\mu: \mathbb{C} \times A \to \mathbb{C}$ be a function with the following properties: for any fixed w situated in the domain $A \subset \mathbb{C}$, $\operatorname{spt}\mu(\ ,w) \subset \Delta^*$, $\mu(\ ,w)$ is measurable in $\mathbb{C}$ and $\|\mu(\ ,w)\|_\infty < 1$, whereas for any fixed $z \in \mathbb{C}$ the function $w \mapsto \mu(z,w)$ is holomorphic in A. Then we have as an immediate consequence of Theorem 18:

THEOREM 19. If $\mu(\ ,w)$ satisfies the above stated conditions, then the function $w \mapsto f(z,w)$, where $f(\ ,w) \in S^\infty(Q)$, being generated by the complex dilatation $\mu(\ ,w)$, is holomorphic (in A) for any finite, fixed z. Moreover, the functions $w \mapsto f_z^{(n)}(z,w)$ are holomorphic (in A) for any fixed $z \in \operatorname{int}\Delta$, $n = 1,2,\ldots$

P r o o f. Let us define $\varkappa(z,w) = (z/\bar{z})^2 \mu(1/z, w)$, where μ is given and satisfies the above stated assumptions. Then $\varkappa(\ ,w)$ is a measurable function with

$$\operatorname{spt} \varkappa(\ ,w) \subset \Delta, \quad \|\varkappa(\ ,w)\|_\infty \leq (Q-1)/(Q+1),$$

and $\varkappa(z,\ )$ is holomorphic in A. Thus $\varkappa(\ ,w)$ generates a unique function $g(\ ,w) \in \Sigma(Q)$ as in the proof of Theorem 18. Consider the function

$$(21.18) \quad f(z,w) = 1/[g(1/z, w) - g(0,w)].$$

It is easily verified that $f(\ ,w) \in S^\infty(Q)$ and the complex dilatation of $f(\ ,w)$ is equal to $\mu(\ ,w)$. Hence the formula (21.18) yields the unique function $f(\ ,w) \in S^\infty(Q)$, generated by $\mu(\ ,w)$. Now $g(z,\ )$ and $g(0,\ )$ are, in view of Theorem 18, holomorphic functions (in A). Moreover, $g(1/z, w) \neq g(0,w)$ for any finite z, and therefore, by (21.18), $f(z,\ )$ is holomorphic. For $z \in \operatorname{int}\Delta$ we have

$$(21.19) \quad f(z,w) = z + \sum_{n=2}^{+\infty} a_n(w) z^n, \quad w \in A.$$

From the identity

$$[1 - g(0,w)z + b_1(z)z^2 + b_2(w)z^3 + \ldots] \times [1 + a_2(w)z + a_3(w)z^2 + \ldots] = 1$$

which is equivalent to (21.18), we can evaluate $a_j(w)$ as polynomials in $g(0,w)$ and $b_k(w)$, $k \leq j-2$; e.g. $a_2(w) = g(0,w)$, $a_3(w) = g(0,w) - b_1(w)$, and so on. Since $g(0,\ )$ and b_j are holomorphic (in A) by Theorem 18 and Corollary 20, so are a_j, $j = 2,3,\ldots$ The proof of holomorphy of $f_z^{(n)}(z,\ )$ is a mere repetition of the proof of Corollary 20.

Theorems 18 and 19 form the basis for Lehto's Majorant Principle which is treated in the next section. A natural question arises whether analogous results are valid for qc mappings of a fixed Jordan domain onto another fixed Jordan domain, e.g. for mappings of the class S_Q. The answer is (cf. Theorem 21) that the mappings of this kind, generated by a complex dilatation $\mu(\ ,w)$ depending holomorphically on w, depend analytically on w only in exceptional cases. It happens to be true only if, for the generated mapping $f(\ ,w)$, the functions $w \mapsto f(z,w)$ are constant for any z on the boundary of the given Jordan domain.

In what follows we assume that D is a fixed Jordan domain in the extended plane $\mathbb{E}$ and the complex parameter w varies in the unit disc $\operatorname{int}\Delta$. We shall need two lemmas.

LEMMA 19. *Suppose that, for any fixed* $w \in \operatorname{int}\Delta$, $z \mapsto f(z,w)$ *is a continuous mapping of a closed Jordan domain* D *into a closed Jordan domain* D'. *Suppose, moreover, that* $w \mapsto f(z,w)$ *is holomorphic in* $\operatorname{int}\Delta$ *for every* $z \in E \subset D$. *Then* $w \mapsto f(z,w)$ *is also holomorphic for any* $z \in \operatorname{cl} E$.

Proof. If E is closed, there is nothing to prove. Suppose that $s \in \operatorname{cl} E \setminus E$ and let (z_n) be a sequence of points in E with $\lim z_n = s$. Since the values of all $f(z,\ )$ omit a fixed disc, the family $\{f(z,\ )\}$ is normal and hence $(f(z_n,\ ))$ contains a subsequence almost uniformly convergent in $\operatorname{int}\Delta$. By continuity of $z \mapsto f(z,w)$ the limiting function is $f(s,\ )$ and hence $f(s,\ )$ is holomorphic.

LEMMA 20. *Suppose that for any fixed* $w \in \operatorname{int}\Delta$, $z \mapsto f(z,w)$ *is a continuous mapping of* D *into* D'. *If* $w \mapsto f(s,w)$ *is holomorphic for some* $s \in \operatorname{fr} D$ *and there exists a* $w_0 \in \operatorname{int}\Delta$ *such that* $f(s,w_0) \in \operatorname{fr} D'$, *then* $w \mapsto f(s,w)$ *is constant in* $\operatorname{int}\Delta$.

Proof. Suppose that $w \mapsto f(s,w)$ is not constant. Then the image domain of $\operatorname{int}\Delta$ is an open set contained in D'. However, $f(s,w_0) \in \operatorname{fr} D'$ and therefore it cannot be an interior point of the image set, and this is a contradiction.

In what follows let μ be a complex-valued function on $\operatorname{int} D \times \operatorname{int}\Delta$ which is measurable in $\operatorname{int} D$ for all fixed $w \in \operatorname{int}\Delta$ and satisfies $\|\mu(\ ,w)\|_\infty < 1$. By the corresponding existence theorem (cf. Section 9) there exists a qc mapping $z \mapsto f(z,w)$ of $\operatorname{int} D$ onto $\operatorname{int} D$ with complex dilatation $\mu(\ ,w)$ admitting a homeomorphic extension on D. We now prove that holomorphy of $f(z,\ )$ implies holomorphy of $\mu(z,\ )$. Namely, we have (Lehto [7]):

THEOREM 20. *Let* $w \mapsto \mu(z,w)$ *be differentiable in* $\operatorname{int}\Delta$ *for a.e.* $z \in D$. *If* $w \mapsto f(z,w)$ *is holomorphic in* $\operatorname{int}\Delta$ *for any* $z \in \operatorname{int} D$, *then* $w \mapsto \mu(z,w)$ *is also holomorphic in* $\operatorname{int}\Delta$ *for a.e.* $z \in D$.

Proof. All $z \mapsto f(z,w)$ map D onto D'. By Lemma 19, $f(s,\ )$ is holomorphic for any $s \in \operatorname{fr} D$ and therefore, by Lemma 20, $w \mapsto f(z,w)$ is constant for any $z \in \operatorname{fr} D$. Thus, given one mapping $f(\ ,w)$, all the others are uniquely determined by their complex dilatation $\mu(\ ,w)$;

cf. Section 9. We may assume without loss of generality that $D = D' = \Delta$. Since $w \mapsto \mu(z,w)$ is differentiable, so are $w \mapsto f_z(z,w)$, $w \mapsto f_{\bar{z}}(z,w)$ (cf. Ahlfors and Bers [1]) for a.e. $z \in D$. Moreover, $f_{z\bar{w}} = f_{\bar{w}z}$, $f_{\bar{z}\bar{w}} = f_{\bar{w}\bar{z}}$. Since $f(z,\ )$ is holomorphic, we have $f_{\bar{w}}(z,w) = 0$ in int Δ. Hence

$$\mu_{\bar{w}}(z,w) = [f_{\bar{z}}(z,w)/f_z(z,w)]_{\bar{w}} = [f_{\bar{z}\bar{w}}(z,w)\, f_z(z,w)$$

$$- f_{\bar{z}}(z,w)\, f_{z\bar{w}}(z,w)][f_z(z,w)]^{-2} = 0 \text{ in int } \Delta \text{ for a.e. } z \in D.$$

Since $\mu(z,\ )$ is differentiable in int Δ, this implies that it is holomorphic in int Δ for a.e. $z \in D$, as desired.

It is easy to give examples showing that $\mu(z,\ )$ holomorphic does not necessarily generate holomorphic $f(z,\ )$. E.g.

$$f(z,w) = [(1+\bar{w})(z+w\bar{z}) + i(w-\bar{w})]/(1-w\bar{w})$$

has complex dilatation $\mu(z,w) = w$ and maps for $|w| \le q < 1$ quasiconformally the upper half-plane $\{z:\ \mathrm{im}\, z > 0\}$ onto itself. However, $f_{\bar{w}} = 0$ for $z = i$ only. We can, however, prove a kind of converse of Theorem 18 (Lehto [7]):

<u>THEOREM 21</u>. <u>Suppose that</u> $w \mapsto \mu(z,w)$ <u>is holomorphic in</u> int Δ <u>for a.e.</u> $z \in \mathrm{int}\, D$. <u>The generated mapping</u> $(z,w) \mapsto f(z,w)$ <u>is holomorphic in</u> $w \in \mathrm{int}\, \Delta$ <u>for any</u> $z \in \mathrm{int}\, D$ <u>iff</u> $f(z,\ )$ <u>is constant in</u> int Δ <u>for any</u> $z \in \mathrm{fr}\, D$.

<u>Proof</u>. The necessity is an immediate consequence of Lemmas 19 and 20. Suppose now that $w \mapsto f(z,w)$ is constant for every $z \in \mathrm{fr}\, D$. We may again assume that $A = A' = \Delta$. Then $z \mapsto f(z,0)$ has a qc extension on $\mathbb{E}$ whose restriction to Δ^* is denoted by g. Put now $h(z,w) = f(z,w)$ for $z \in \Lambda$ and $h(z,w) = g(z)$ for $z \in \Delta^*$. Obviously $h(\ ,w)$ is a qc selfmapping of $\mathbb{E}$ whose complex dilatation is equal to $\mu(\ ,w)$ in int Λ and to $g_{\bar{z}}/g_z$ a.e. in Λ^*, and hence holomorphic as a function of w for a.e. z. As shown by Lehto [6], any selfmapping $f(\ ,w)$ of $\mathbb{E}$ generated by a complex dilatation $\mu(\ ,w)$ such that the corresponding $\mu(\ ,w)$ are holomorphic a.e. in int Λ and $f(z_j,w) = c_j$, $j = 1, 2, 3$, where z_j, c_j do not depend on w, yields a function $w \mapsto f(z,w)$ holomorphic in int Λ for any fixed, finite z. In our case this condition is satisfied with z_j being three arbitrarily chosen points on fr D. This proves the sufficiency.

We conclude this section by giving two examples of mappings of the class S_Q depending analytically on w.

<u>Example 1</u>. Let $f(z,w) = z|z|^w$, $z \in \Delta$. Hence $\mu(z,w) = [w/(z+w)]$

$\varkappa(z/\bar{z})$. If $|w| \le q < 1$, then $|\mu(z,w)| \le q$ so that $f(\ ,w) \subset S_Q$ with $Q = (1+q)/(1-q)$. Obviously $f(e^{i\theta}, w) = e^{i\theta}$ for any $|w| \le q$.

Example 2. Let

$$f(z,w) = 2|z|^2[\bar{z} + wz + \sqrt{(\bar{z}+wz)^2 - 4w|z|^4}\,]^{-1}, \quad z \in \Lambda.$$

Hence $\mu(z,w) = w(z/\bar{z})^2$. If $|w| \le q < 1$, then $|\mu(z,w)| \le q$ so that $f(\ ,w) \in S_Q$ with $Q = (1+q)/(1-q)$.

22. Lehto's Majorant Principle and its applications

Holomorphic dependence of functions f of the class $\Sigma(Q)$ (or $S^{\infty}(Q)$, respectively) on the complex parameter w, in case of the complex dilatation μ generating f as a holomorphic function of w, permits us to apply various theorems of the classical function theory while investigating such mappings. This fact leads to a general principle formulated by Lehto [7] that enables us to obtain new results for functions of the class $\Sigma(Q)$ (or $S^{\infty}(Q)$, respectively) being counterparts of classical results related to functions of the class Σ (or S, respectively) and to obtain new, simple proofs of many results recently established for the classes $\Sigma(Q)$ and $S^{\infty}(Q)$ by various authors. In what follows we always assume that all g in Σ have the hydrodynamic normalization: $g(z) - z \to 0$ as $z \to \infty$.

It is well known that for $g \in \Sigma$ we have

(22.1) $\quad |g(z) - z| < 3/|z|, \quad |z| > 1$

(cf. e.g. Pólya and Szegö [1], vol. II, p. 25, Problem 144). Let z_j, $j = 1, 2, 3$, be the vertices of an equilateral triangle inscribed in the circle $\{z: |z| = 3\}$. Then from (22.1) it follows that the values of $g(z_j)$, $g \in \Sigma$, are situated in three disjoint discs $\Delta^1(z_j)$. Since $|z_j - z_k| = 3\sqrt{2} > 2$ for $j \ne k$, there exists $d > 0$ such that the spherical distance between $g(z_j)$ and $g(z_k)$ is greater than d for $j, k = 1, 2, 3$, $j \ne k$. Hence Σ and $\Sigma(Q)$ are normal families (with respect to the convergence in the spherical metric); cf. Lehto and Virtanen [1], p. 76, or [2], p. 73.

If the value w is omitted by $g \in \Sigma$, then $|w| \le 2$ (cf. Pólya and Szegö [1], vol. II, p. 25, Problem 139). This implies that $|g(0)| \le 2$ for any $g \in \Sigma(Q)$ with arbitrary $Q \ge 1$. Now, with any $f \in S^{\infty}(Q)$ we can associate a unique $g \in \Sigma(Q)$ such that

$$f(z) = 1/[g(1/z) - g(0)], \quad |z| < 1.$$

Therefore $S^\infty(Q)$ is a normal family as well.

Let Φ be a complex-valued functional defined in all classes $\Sigma(Q)$. We say that Φ is continuous, if the uniform convergence $g_n \rightrightarrows g$ (in spherical metric) implies $\Phi[g_n] \to \Phi[g]$. Since $\Sigma(Q)$ are closed normal families, for any continuous functional Φ there exists in each $\Sigma(Q)$ an extremal function maximizing $|\Phi[g]|$ in $\Sigma(Q)$. Put

$$(22.2)\quad M(q) = \max_{g \in \Sigma(Q)} |\Phi[g]|, \quad q = \frac{Q-1}{Q+1}.$$

Obviously M is an increasing (not necessarily strictly increasing) function of q in $[0; 1]$.

LEMMA 21. *If Φ is a continuous functional defined on each $\Sigma(Q)$, then the function M given by (22.2) is continuous in $[0; 1)$.*

Proof. Suppose that q_0 is an arbitrary number in $(0; 1)$. Because $M(q)$ is increasing, we have

$$(22.3)\quad \lim_{q \to q_0-} M(q) \le M(q_0) \le \lim_{q \to q_0+} M(q).$$

Suppose that $q \le q_0$. Let g_0 be an extremal function for $|\Phi|$ in $\Sigma(Q_0)$ with complex dilatation μ. Let g_q be the function in $\Sigma(Q)$ generated by complex dilatation (q/q_0). Since $\{g_q\}$ is a normal family, there exists an increasing sequence (q_n) such that $\lim q_n = q_0$ and the mappings g_q, $q = q_n$, converge uniformly in spherical metric to a limiting function g. Then the complex dilatation of g is equal μ (cf. Lehto and Virtanen [1], p. 196, or [2], p. 187) and, by the uniqueness theorem, $g = g_0$. In view of continuity of Φ, we have $\lim \Phi[g_{q_n}] = \Phi[g_0]$. Thus

$$M(q_0) = |\Phi[g_0]| = \lim |\Phi[g_{q_n}]| \le \lim M(q_n).$$

Since $q_n \to q_0$, we have

$$M(q_0) \le \lim_{q \to q_0-} M(q)$$

and comparing this with (22.3) we obtain left-hand side continuity at q_0.

Take now q, $q_0 < q < 1$, and let g denote an extremal mapping in $\Sigma(Q)$. Choose a decreasing sequence $q_n \to q_0+$ so that g_q, $q = q_n$, converge uniformly to g. Obviously $g \in \Sigma(Q_0)$ (cf. Lehto and Virtanen [1], p. 76, or [2], p. 74) and this implies

$$M(q_0) \geq |\Phi[g]| = \lim |\Phi[g_{q_n}]| = \lim M(q_n) = \lim_{q \to q_0+} M(q).$$

A comparison with (22.3) yields continuity to the right at q_0. This part of the proof also applies for the case $q_0 = 0$. If Φ makes sense for Σ, then using the fact that $g \in \Sigma$ implies

$$g \circ h/(1+\tfrac{1}{n}) \in \Sigma(Q), \quad h(z) = (1+\tfrac{1}{n})z,$$

for some $q = q(n)$ which tends to 1 as $n \to +\infty$, we can prove in a similar way the left continuity of $M(q)$ at $q = 1$.

Suppose that Ψ is an analytic function of N complex variables

$$w_1, \ldots, w_N, \quad N = m + \sum_{k=0}^{s} p_k,$$

and that $\Phi[g]$, $g \in \Sigma(Q)$, arises by replacing m variables w_j in Ψ by the first m Laurent coefficients of g at $z = \infty$, then p_0 variables w_j by the values of g at p_0 points

$$z_{0,1}, z_{0,2}, \ldots, z_{0,p_0},$$

then p_1 variables w_j by the values of g' at p_1 points

$$z_{1,1}, \ldots, z_{1,p_1},$$

and finally the remaining p_s variables w_j by the values of $g^{(s)}$ at p_s points

$$z_{s,1}, \ldots, z_{s,p_s}.$$

Obviously $z_{0,k}$ can be chosen arbitrarily, whereas the remaining points $z_{j,k}$ must be taken from Δ^*. Evidently we must in general case confine ourselves to a subclass $\Sigma(Q)$ for which the substitution makes sense. A functional obtained in this way is called _analytic_. Analytic functionals are obviously continuous.

We are going to prove the following _Lehto's Majorant Principle_ (Lehto [4, 6]):

THEOREM 22. _If Φ is an analytic functional which is defined in every $\Sigma(Q)$ and vanishes for the identity mapping_ id, _then $M(q)/q$ is increasing on the interval $(0; 1)$; M being defined by_ (22.2).

Proof. Choose q and q' so that $0 < q < q' < 1$ and take an arbitrary $g_1 \in \Sigma(Q)$, $Q = (1+q)/(1-q)$. Let μ be the complex dilatation of g_1. Let $f \in \Sigma(Q')$, $Q' = (1+q')/(1-q')$, be generated by the complex dilatation $w\mu$, where $|w| \leq q'/q$. By Theorem 18 and Corollaries 19 and 20 the functional $g \mapsto \Phi[g]$ depends analytically on w in the disc $\{w: |w| < q'/q\}$. If $w = 0$, then $g = \mathrm{id}$ and therefore $\Phi[g] = 0$ at $w = 0$. Therefore, by the Schwarz lemma,

$$|\Phi[g]| \leq (q/q')\, M(q')\, |w|.$$

For $w = 1$ we have $g = g_1$. Since $g_1 \in \Sigma(Q)$ was chosen in an arbitrary manner, we can take the extremal mapping as g_1 so that

$$M(q) \leq (q/q')\, M(q')$$

or

(22.4) $\quad (1/q)\, M(q) \leq (1/q')\, M(q'), \quad 0 \leq q < q' < 1,$

and this proves the monotoneity of $q \mapsto M(q)/q$.

In what follows we denote $\max|\Phi[g]|$ in Σ by $M(1)$. However, not all Φ defined for every $\Sigma(Q)$ make sense for Σ. E.g. $\Phi[g] = g(0)$ is not defined for $g \in \Sigma$. However, if $M(1)$ makes sense, then, by Lemma 21,

$$M(1) = \lim_{q \to 1-} M(q).$$

Therefore we can define in any case $M(1)$ as $\lim_{q \to 1-} M(q)$.

It can easily be observed that Theorem 22 remains valid for functionals analytic in the following more general sense. Let Φ be a complex-valued functional defined for the class Σ. We call Φ <u>analytic</u> if for every function $(z,w) \mapsto g(z.w)$, analytic in the domain $\Lambda^* \times A$ and satisfying the property $g(\ ,w) \in \Sigma$ for any fixed $w \in A$, $w \mapsto \Phi[s]$, $s = g(\ ,w)$, $w \in A$, is an analytic function. The same remark concerns further results.

<u>COROLLARY 21</u>. <u>If Φ is an analytic functional in Σ and $\Phi[\mathrm{id}] = 0$, then</u>

(22.5) $\quad M(q) \leq q\, M(1).$

<u>If $\Phi[\mathrm{id}] \neq 0$, then</u>

(22.6) $\quad M(q) \leq [q\, M(1) + M(0)]/[1 + q\, M(0)/M(1)].$

<u>Proof</u>. The inequality (22.5) is an immediate consequence of (22.4) and Lemma 21. In order to prove (22.6) consider $g_1 \in \Sigma(Q)$ with

complex dilatation μ. Let g be generated by the complex dilatation $w\mu$, where $|w| < 1/q$, $q = (Q-1)/(Q+1)$. Then $\Phi[g]$ is a holomorphic function of w in the disc $\{w: |w| < 1/q\}$ that takes the value $M(0)$ at $w = 0$ and is bounded by $M(1)$ there. Hence

$$[M(1)(\Phi[g] - M(0))]/[M^2(1) - M(0)\,\Phi[g]]$$

satisfies the assumptions of Schwarz's lemma (by taking posisibly $e^{i\alpha} \times \Phi[g]$ we may assume that $\Phi[\mathrm{id}] > 0$). Thus

$$(22.7) \quad M(1)\{|\Phi[g]| - M(0)\}/\{M^2(1) - M(0)|\Phi[g]|\} \le q|w|.$$

For $w = 1$ we obtain $g = g_0$ and since g_0 is arbitrary, (22.6) readily follows from (22.7).

COROLLARY 22. *If equality holds in* (22.5) *for some* q, *then it holds for all values* q.

Proof. Suppose that $M(q) = q\,M(1)$ for some $q < 1$. Let g_1 be extremal and let μ be its complex dilatation. For functions with complex dilatation $w\mu$, $|w| < 1/q$, the Schwarz lemma gives $|\Phi[g]| \le q|w| \times M(1)$ with equality in an interior point $w = 1$ so that $|\Phi[g]| = q|w| \times M(1)$ in the whole disc $\{w: |w| < 1/q\}$. If $q' \in (0;1)$ is arbitrary then $w = q'/q$ yields a function $\tilde{g}$ in $\Sigma(Q')$, $Q' = (1+q')/(1-q')$, so that $|\Phi[\tilde{g}]| = q'M(1)$. By (22.5) $\tilde{g}$ is extremal in $\Sigma(Q')$. In fact, in this case the sign of equality in (22.5) holds for every $q \in (0;1)$.

It is easily seen that Theorem 22 and its corollaries have counterparts for the class $S^{\infty}(Q)$. Since their reformulation is obvious and the proofs are identical, we would not quote these results here.

If Φ is an analytic functional and g depends analytically on a parameter w, then $\mathrm{re}\,\Phi[g]$ is a harmonic function. Lehto's Majorant Principle can be stated also for functionals of this type. Suppose that

$$(22.8) \quad m(q) = \min_{g \in \Sigma(Q)} \mathrm{re}\,\Phi[g], \quad q = \frac{Q-1}{Q+1},$$

and put:

$$m(1) = \min_{g \in \Sigma} \mathrm{re}\,\Phi[g],$$

respectively

$$m(1) = \lim_{q \to 1-} m(q)$$

if $\mathrm{re}\,\Phi[g]$ makes no sense for $g \in \Sigma$, and

$$m(0) = \mathrm{re}\,\Phi[\mathrm{id}].$$

Then we have the following unpublished result of O. Lehto (cf. Göktürk [1]):

THEOREM 23. *Let* Φ *be an analytic functional defined in every* $\Sigma(Q)$. *Then, for every* $g \in \Sigma(Q)$,

$$(22.9) \qquad (1 - \tfrac{1}{Q})[m(1) - m(0)] \le \operatorname{re} \Phi[g] - m(0) \le (Q-1)[m(0) - m(1)];$$

m *being defined by* (22.8).

Proof. Let g_1 be an arbitrary mapping in $\Sigma(Q)$, $1 \le Q < +\infty$, with complex dilatation μ. Let g be a function of the class Σ generated by the complex dilatation w, $|w| < 1/q$, $q = (Q-1)/(Q+1)$. By Theorem 18 $u = \operatorname{re} \Phi[g]$ is a harmonic function of w for $|w| < 1/q$. By Poisson's formula, we have

$$(22.10) \qquad u(w) - m(1) = \frac{1}{2\pi} \int_0^{2\pi} \frac{\rho^2 - r^2}{|\rho e^{i\theta} - w|^2} [u(\rho e^{i\theta}) - m(1)] d\theta,$$

where $|w| = r < \rho < 1/q$. Since $u(w) - m(1) \ge 0$ and

$$(22.11) \qquad \frac{\rho - r}{\rho + r} \le \frac{\rho^2 - r^2}{|\rho e^{i\theta} - w|^2} \le \frac{\rho + r}{\rho - r},$$

formula (22.10) yields

$$u(w) - m(1) \le \frac{1}{2\pi} \frac{\rho + r}{\rho - r} \int_0^{2\pi} [u(\rho e^{i\theta}) - m(1)] d\theta.$$

The arithmetic mean of $u(\rho e^{i\theta}) - m(1)$ over $[0; 2\pi]$ is equal $u(0) - m(1) = m(0) - m(1)$ so that

$$u(w) - m(1) \le \frac{\rho + r}{\rho - r} [m(0) - m(1)].$$

If $\rho \to 1/q$, we obtain

$$u(w) - m(1) \le \frac{1 + qr}{1 - qr} [m(0) - m(1)],$$

or

$$(22.12) \qquad u(w) - m(0) \le \frac{2qr}{1 - qr} [m(0) - m(1)].$$

If we take $w = 1$, then $r = 1$ and $u = \operatorname{re} \Phi[g]$, so that (22.12) becomes the right inequality in (22.9). Similarly, the left inequality in (22.11) and the formula (22.10) imply the left inequality in (22.9).

The inequality (22.9) immediately yields

COROLLARY 23. *We have*

$$(22.13)\quad \frac{1-q}{1+q}m(0)+\frac{2q}{1+q}m(1)\le m(q) \quad \text{for} \quad 0\le q<1.$$

Moreover, if $\tilde{M}(q)=\max_{g\ (Q)} \operatorname{re}\Phi[g]$, $q=(Q-1)/(Q+1)$, *then we have*

$$(22.14)\quad \tilde{M}(q)\le \frac{1+q}{1-q}m(0)-\frac{2q}{1-q}m(1).$$

COROLLARY 24. *Theorem* 23 *and Corollary* 23 *also hold if*

$$m(q)=\min_{f\in S^\infty(Q)}\Phi[f],\quad q=\frac{Q-1}{Q+1}.$$

We now give several examples being applications of Lehto's Majorant Principle. We put everywhere $q=(Q-1)/(Q+1)$.

Example 3. Suppose that

$$g(z)=z+b_1z^{-1}+b_2z^{-2}+\dots,\quad g\in\Sigma(Q),$$

and $\Phi[g]=b_n$. Then $\Phi[id]=0$ and hence, by (22.5),

$$|b_n|\le q\max|b_n|,\quad n=1,2,\dots$$

Therefore $|b_1|\le q$ and $|b_2|\le \frac{2}{3}q$. The latter bounds are sharp. The extremal functions are

$$g(z)=\begin{cases} z+q/z, & |z|>1\\ z+q\bar{z}, & |z|\le 1\end{cases}\quad (n=1)$$

and

$$g(z)=\begin{cases} (z^{3/2}+qz^{-3/2})^{2/3}, & |z|>1\\ (z^{3/2}+q\,\bar{z}^{3/2})^{2/3}, & |z|\le 1\end{cases}\quad (n=2).$$

Example 4. Suppose that $f(z)=z+a_2z^2+\dots$, $f\in S^\infty(Q)$, and $\Phi[f]=a_2$. Then $\Phi[id]=0$, $M(1)=2$ and hence, by (22.5), or rather its counterpart for $S^\infty(Q)$, we have $|a_2|\le 2q$. This result is sharp and the extremal function is

$$f(z)=\begin{cases} z(1-qz)^{-2}, & |z|<1,\\ z\bar{z}(\bar{z}-2q|z|+q^2z)^{-1}, & |z|\ge 1.\end{cases}$$

Similarly, if $\Phi[f]=a_2^2-a_3$, then $\Phi[id]=0$, $M(1)=1$, so that by (22.5) we have

$$|a_2^2 - a_3| \leq q$$

for every $f \in S^\infty(Q)$. Also this result is sharp and the extremal function in $S^\infty(Q)$ is

$$f(z) = \begin{cases} z(1+qz^2)^{-1}, & |z|<1, \\ z\bar{z}(\bar{z}+qz)^{-1}, & |z| \geq 1. \end{cases}$$

Both results in this example are due to Kühnau [2].

Example 5. With any $g \in \Sigma$ we can associate an infinite Grunsky matrix by means of the equation

$$\log \frac{g(z)-g(s)}{z-s} = -\sum_{m=1}^{+\infty} \sum_{n=1}^{+\infty} b_{m,n} z^{-m} s^{-n}.$$

Consider now

$$\Phi[g] = \sum_{m,n=1}^{N} b_{m,n}\lambda_m\lambda_n \quad \text{and} \quad C(\lambda_1, \dots, \lambda_N) = \sum_{m=1}^{N} |\lambda_m|^2/m,$$

where $\lambda_1, \dots, \lambda_n \in \mathbb{C}$. Then $\Phi[g]$ is an analytic functional in Σ which vanishes for $g = \mathrm{id}$. Moreover, by Grunsky's inequality (cf. e.g. Pommerenke [1], p. 60), we have $|\Phi[g]| \leq C(\lambda_1, \dots, \lambda_N)$. Thus, by (22.5), we have for any $g \in \Sigma(Q)$ and any complex $\lambda_1, \dots, \lambda_N$ (Kühnau [3, 5]):

$$|\Phi[g]| \leq q\, C(\lambda_1, \dots, \lambda_N).$$

Example 6. We now evaluate $M(q)$ for $\Phi[g] = g(0)$, $g \in \Sigma(Q)$. It is well known that for $g(z) = z + b_1 z^{-1} + \dots$, $g \in \Sigma$, the values w not taken in Δ^* satisfy $|w| \leq 2$ (cf. e.g. Pólya and Szegö [1], vol. II, p. 25, Problem 139). Hence we have $|g(0)| \leq 2$ for any $g \in \Sigma(Q)$, $1 \leq Q < +\infty$. Thus $M(1) = \lim_{q \to 1-} M(q) \leq 2$ and, by (22.5),

$$M(q) \leq 2q.$$

We can easily show that this bound is sharp. To this end consider

$$g(z) = \begin{cases} z + q^2/z, & |z| > 1, \\ z + 2q(|z|-1) + q^2\bar{z}, & |z| \leq 1. \end{cases}$$

It is readily verified that g is extremal for Φ in $\Sigma(Q)$. In this case Φ is not defined in Σ.

Example 7. Consider now the functional $f \mapsto \operatorname{re}\log[f(z)/z]$ in $S^\infty(Q)$. With the notation of Theorem 23 we have $m(0) = 0$, $m(1) = -2\log(1+r)$, where $r = |z|$. Hence, by (22.9), $r(1+r)^{-4q/(1+q)} \leq |f(z)|$

and with $r=1$ we obtain the value of Koebe's constant $(\frac{1}{4})^{2q/(1+q)}$ for $S^\infty(Q)$. On the other hand, the right-hand side inequality in (22.9) gives $|f(z)| \le r(1+r)^{4q/(1-q)}$. Thus the image curve of $\{z: |z|=1\}$ under $f \in S^\infty(Q)$ is contained in the annulus determined by the inequalities

$$(\frac{1}{4})^{2q/(1+q)} \le |w| \le 4^{2q/(1-q)}.$$

These bounds are due to Göktürk [1] (cf. also Tanaka [1]). The exact values of

$$\sup_{f \in S^\infty(Q)} \max_{|z|=1} |f(z)| \quad \text{and} \quad \inf_{f \in S^\infty(Q)} \min_{|z|=1} |f(z)|$$

are calculated by Kühnau [10].

Example 8. We now derive the estimate for the deviation of $f \in S^\infty(Q)$ from the identity in Δ. If $g \in \Sigma(\infty)$, then (cf. (22.1)):

$$|g(s)-s| < 3/|s|, \quad |s|>1.$$

Hence, by the Majorant Principle,

$$|g(s)-s| \le 3q/|s|, \quad |s| \ge 1, \quad g \in \Sigma(Q).$$

Obviously, each $f \in S^\infty(Q)$ can be written in the form

$$f(z) = 1/[g(s)-g(0)], \quad g \in \Sigma(Q), \quad s = 1/z,$$

so that

$$|f(z)-z| = \frac{s-g(s)+g(0)}{s[g(s)-g(0)]},$$

and by Examples 6 and 7 we obtain

$$|f(z)-z| \le |z||f(z)|(3q|z|+2q)$$
$$\le 2qr^2(1+\frac{3}{2}r)(1+r)^{4q/(1-q)}, \quad r=|z|.$$

This estimate is sharp for $q=0$ and also asymptotically sharp for fixed q and $r \to 0+$.

Example 9. Consider $\Phi[f] = \{f, \} = (f''/f')' - \frac{1}{2}(f''/f')^2$, i.e. $\Phi[f]$ denotes the Schwarzian derivative of $f \in S^\infty(Q)$. It is well known that $|\{f,z\}| \le 6(1-|z|^2)^{-2}$ for any $f \in S$. The Majorant Principle gives at once $|\{f,z\}| \le 6q(1-|z|^2)^{-2}$ for $f \in S^\infty(Q)$ which was obtained by Kühnau [2]. The above estimates are precise.

III. A REVIEW OF VARIATIONAL METHODS AND BASIC APPLICATIONS IN ELECTRICAL ENGINEERING

23. Belinskiĭ's variational method

Variational methods have been in use since long time in various papers concerned with extremal problems in the theory of qc mappings. Already Teichmüller [1] and Ahlfors [2] applied an ad hoc variational method (cf. also Ahlfors and Bers [1], Bers [4], Cheng Bao-long [1], and Kruškal [5, 8]). More systematic approaches are due to Belinskiĭ [1, 2], Schiffer [2], and Renelt [1, 3].

For variational methods it is essential for us to be able to construct, given an element f belonging to the class of mappings in question, a uniparametric family of comparison functions of the form $\tilde{f} = f + t\,h + o(t)$ belonging to the same class for all sufficiently small numbers $t > 0$.

Let F be a real-valued functional defined for all complex-valued mappings of the class C^1 of a domain D, $D \subset \mathbb{C}$. This functional is said to be <u>differentiable in the sense of Gâteaux</u> at the element f, if for $f + h \in D_{\mathsf{F}}$ (the domain of F), $t \to 0+$, we have

$$\mathsf{F}[f + th] - \mathsf{F}[f] = \operatorname{re}(t\,\partial\mathsf{F}[f][h]) + o(t), \tag{23.1}$$

where $o(t)/t \to 0$ uniformly on compact subsets of D and $\partial\mathsf{F}[f]$ is a linear functional. The functional $\partial\mathsf{F}[f]$ is called <u>the first Gâteaux variation</u> (or: <u>derivative</u>) of F at f.

It can be easily seen that if F attains an extremum w.r.t. a subclass of the class in question for a mapping f and we know the family of comparison functions of this subclass in the form $f + t\,h + o(t)$, then $\partial\mathsf{F}[f][h] = 0_D$. Thus this is a necessary condition for an extremum to exist.

Belinskiĭ [1, 2] proposed a variational method based on considering a mapping g defined as follows:

$$g(w) = w + \varepsilon\,(\bar{w} - \bar{w}_0) \quad \text{for} \quad |w - w_0| < r,$$

$$g(w) = w + \varepsilon r^2/(w - w_0) \quad \text{for} \quad |w - w_0| > r,$$

where $|\varepsilon| < 1$. Thus inside $\Delta^r(w_0)$ it is a qc mapping with constant complex dilatation ε, and outside $\Delta^r(w_0)$ it is a conformal one. Besides g is a homeomorphism of $\mathbb{E}$. If we consider the composed mapping $g \circ f$, where f is qc, then for sufficiently small numbers ε we get qc mappings with maximal dilatation increased arbitrarily little, so this gives a way to form a family of comparison functions. Even such a simple variational function may be sometimes useful. When considering, however, extremal problems e.g. within the class S_Q^* it is better to choose a function g which possesses a small, but not necessarily constant complex dilatation, and keeps the points 0, 1, ∞ invariant.

We have the following theorem of Kruškal [4, 7] which extends a result of Belinskiĭ [1] in the sense of weakening the hypotheses:

THEOREM 24. *Suppose that* $\mu_0 : \mathbb{C} \to \mathbb{C}$ *is a measurable function with compact support contained in a domain* D *and such that* $\|\ _0\| < 1$. *Then for any sufficiently small number* $T > 0$ *and* $t \in (0; T)$ *there is a qc mapping* f^* *of the class* S_Q^* *with* Q *given by* (9.2) *and* $\mu = t\mu_0$, *of the form*

$$(23.2) \qquad f^*(w) = w + t\frac{w(1-w)}{\pi} \iint_D \frac{\mu_0(z)\,dxdy}{z(1-z)(w-z)} + \mathcal{O}(t^2),$$

where $\mathcal{O}(t^2)$ *is a function which after dividing by* t^2 *remains bounded for* $t \to 0+$ *on an arbitrary compact subset of* D. *Moreover, the complex dilatation* μ *of* f^* *satisfies the condition*

$$(23.3) \qquad \|\mu(1 - f_z^{*\prime})\|_p < Mt^2,$$

where M is a constant independent of t and p, $p > 2$, depends on $\|\mu_0\|_\infty$ *only*.

Proof (the original proof of J. Krzyż). Let $f: \mathbb{E} \to \mathbb{E}$ be a homeomorphic solution of (4.2), where $\mu = t\mu_0$, having for $w \to \infty$ the form $f(w) = w + o(1)$. Then, by Lemma 4, we have (cf. the proof of Lemma 11):

$$f(w) = w - \frac{1}{\pi} \iint_D \frac{f_{\bar z}'(z)}{z - w}\,dxdy = w - \frac{1}{\pi}\iint_D \frac{\mu(z)\, f_z'(z)}{z - w}\,dxdy.$$

Set $g(w) = f(w) - f(0)$ which does not affect f_w and μ as well. Hence

$$(23.4) \qquad g(w) = w[1 - (1/\pi)\iint_D \mu(z)\, f_z'(z)\, z^{-1}(z-w)^{-1}dxdy].$$

Let $f^*(w) = g(w)$ with λ chosen so that $f^*(1) = 1$, i.e., by $f^{*\prime}_w = g'_w$, we have

$$(23.5)\qquad \lambda = 1 + (1/\pi) \iint_D \mu(z)\, f^{*\prime}_z\, z^{-1}(z-1)^{-1} dxdy.$$

Multiplying both sides of (23.4) by λ and taking into account (23.5), we get

$$(23.6)\qquad f^*(w) = w\Big[1 - \frac{w-1}{\pi} \iint_D \frac{\varkappa(z)\,dxdy}{z(z-1)(z-w)}\Big], \quad \text{where } \varkappa = \mu f^{*\prime}_w.$$

By Theorem 1 the function f^*_w is of a class L^p, where p, $p>2$, depends on $\|\mu_0\|_\infty$ only. Since μ_0 and $\varkappa$ vanish outside D, then

$$(23.7)\qquad \|\mu - \varkappa\|_p \le \Big[\iint_D |\mu(1 - f^{*\prime}_w)|^p\, dxdy\Big]^{1/p}$$

$$\le t\|\mu_0\|_\infty \Big[\iint_D |1 - f^{*\prime}_w|^p\, dxdy\Big]^{1/p}.$$

Thus it remains to estimate the last factor. From (23.5) we get

$$(23.8)\qquad |1 - \lambda^{-1}| = \Big|\frac{1}{\pi}\iint_D \frac{\mu(z) f^{*\prime}_z(z)}{z(z-1)}\, dxdy\Big| \le t\|\mu_0\|_\infty \frac{1}{\pi}\iint_D \Big|\frac{f^{*\prime}_z(z)}{z(z-1)}\Big| dxdy$$

$$\le M_1 t,$$

where M_1 is a constant independent on t. Hence, for sufficiently small t, e.g. for $0<t<T$:

$$(23.9)\qquad |\lambda| < (1 - M_1 t)^{-1} \le M_2.$$

In view of Lemma 8, we have $f_w(w) = 1 - (1/\pi)\iint_D (z-w)^{-2}\mu(z) \times g_z(z)\, dxdy$, and hence

$$(23.10)\quad |1 - f^{*\prime}_w(w)| = |1 - \lambda g'_w(w)| = \Big|1 - \lambda + \frac{\lambda}{\pi}\iint_D \frac{\mu(z)\, g'_z(z)}{(z-w)^2}\, dxdy\Big|$$

$$\le |\lambda|\,|1 - \lambda^{-1}| + |\lambda|\,\Big|\frac{1}{\pi}\iint_D \frac{\mu(z)\, g'_z(z)}{(z-w)^2}\, dxdy\Big|$$

a.e. in D. Moreover Lemma 8 yields

$$\Big\|\frac{1}{\pi}\iint_D \frac{\mu(z)\, g'_z(z)}{(z-w)^2}\, dxdy\Big\|_p \le \|S\|_p \|\mu g'_z\|_p = \|S\|_p\, t\|\mu_0 g'_z\|_p = M_3 t.$$

Therefore, by (23.10) and Minkowski's inequality, we have

$$\Big(\iint_D |1 - f_w^*(w)|^p du dv\Big)^{1/p} \le |\lambda|\,|1 - \lambda^{-1}|\,|D|^{1/p} + |\lambda| M_3 t,$$

and hence, by (23.8) and (23.9), we get $\|1 - f_w^{*\prime}\|_p < M_4 t$. Taking into account the last inequality in (23.7), we obtain $\|\mu - \varkappa\|_p < Mt^2$, where $M = M_4 \|\mu_0\|_\infty$, and this proves (23.3).

Let us rewrite the formula (23.6) as

$$f^*(w) = w\Big[1 - \frac{w-1}{\pi}\iint_D \frac{\mu(z)\,dxdy}{z(z-1)(z-w)}\Big] - \frac{w(w-1)}{\pi}\iint_D \frac{(\varkappa - \mu)(z)\,dxdy}{z(z-1)(z-w)}.$$

Applying to the latter term Hölder's inequality, we obtain, by (23.3), the formula (23.2), as desired.

Proceeding along the same lines it is not difficult to obtain variational formulae for qc mappings of a disc, strip or half-plane onto itself. To this end we have to continue the mapping in question onto the whole plane with help of the Schwarz reflection principle. The assumption for μ_0 to vanish outside of a compact set here is not essential (as before, including Theorem 24), but it is cumbersome to get free of it. In particular, it is possible to prove the following result:

<u>Suppose that</u> $\mu_0 : \mathbb{C} \to \mathbb{C}$ <u>is a measurable function such that</u> $\|\mu_0\|_\infty < 1$. <u>Then for any sufficiently small number</u> $T > 0$ <u>and</u> $t \in (0; T)$ <u>there is a qc mapping</u> f^* <u>of the class</u> S_Q <u>with</u> Q <u>given by</u> (9.2) <u>and</u> $\mu = t\mu_0$, <u>of the form</u>

$$(23.11)\quad f^*(w) = w + t\,\frac{w(1-w)}{\pi}\iint_\Delta \Big[\frac{\mu_0(z)}{z(1-z)(w-z)} + \frac{\overline{\mu_0(z)}}{\bar z(1-\bar z)(1-w\bar z)}\Big]dxdy + \mathcal{O}(t^2).$$

<u>Moreover, the complex dilatation</u> μ <u>of</u> f^* <u>satisfies the condition</u> (23.3), <u>where</u> M <u>is a constant independent of</u> t <u>and</u> p, $p > 2$, <u>depends on</u> $\|\mu_0\|_\infty$ <u>only (it is clear that now</u> $\|\ \|_p$ <u>and</u> $\|\ \|_\infty$ <u>are related to</u> Δ <u>instead of</u> $\mathbb{C}$).

With help of the above variational formulae Belinskiĭ [1] derived some estimates within the class S_Q. In particular he established Corollary 15 in a slightly weaker version, since his estimate has the form

$$|f(s) - s| \le M(Q-1), \quad \text{where} \quad M = (1/\pi)\iint_{\mathbb{C}} |z(1-z^2)|^{-1} dxdy \doteq 4.5.$$

The same method, however, enables to get also the estimate of $|f(s) - s|$ by $M \log Q$, and a simple calculation gives $M = (1/4\pi^2)[\Gamma(\frac{1}{4})]^4$

which agrees with Corollary 15.

The method of Belinskiĭ is also described in his book [2] and in the books of Kruškal [7] and [8], who generalized it to Riemann surfaces and applied to various interesting extremal problems in several papers listed in his books. The books also include information about the important papers in this direction written by P. A. Biluta. We remark that the generalizations of Kruškal include normalization of the mappings at an arbitrary finite number of points: $f(z_j) = w_j$, $j = 1, \ldots, n$, and – if the mappings are conformal in a subdomain – also a normalization of the derivatives of an arbitrary finite order at the corresponding points.

24. A simple example of application

We are going to give a simple example of application of Theorem 24 for the classes $E_Q^{r,R}$ introduced in Section 19 (cf. Ławrynowicz [9]).

COROLLARY 25. *Suppose that* $\mu_0: \Delta^r \rightarrow ¢$ *is a measurable function with compact support contained in an annulus* $D = \{w: s < |w| < s + \Delta s\}$, $r < s < 1 - \Delta s$, *and such that* $\|\mu_0\|_\infty < 1$. *Then for any sufficiently small number* $T > 0$ *and* $t \in (0; T)$ *there is a qc mapping* f^* *of a class* $E_Q^{r,r'}$ *with* Q *given by* (9.2) *and* $\mu = t\mu_0$, *of the form*

$$(24.1)\quad f^*(w) = \begin{cases} w\left[1 - t\int_s^{s+\Delta s} 2\mu_0(\rho)\rho^{-1}d\rho\right] + \mathcal{O}(t^2) & \text{for } |w| < s, \\ w + \mathcal{O}(t^2) & \text{for } |w| \geq s, \end{cases}$$

where $\mathcal{O}(t^2)$ *is a function which after dividing by* t^2 *remains bounded for* $t \rightarrow 0+$ *for an arbitrary compact subset of* D. *Moreover, the complex dilatation* μ *of* f^* *fulfills the condition* (23.3), *where* M *is a constant independent of* t *and* p, $p > 2$, *depends on* $\|\mu_0\|_\infty$ *only*.

Proof. Let us continue μ_0 into the inner disc Δ^r by the value 0 and then into the exterior of Δ by the formula $\mu_0(w) = e^{4i \arg w} \times \overline{\mu_0(1/\overline{w})}$. Now we can apply Theorem 24. By Lemma 18, in particular condition (iii), the mapping f^* constructed there, when restricted to Δ^r, belongs to a class $E_Q^{r,r'}$. Hence it remains to show that in our case the formula (23.2) reduces to (24.1).

By Lemma 18, in particular condition (iii), the function $f^*|\Delta_r$, given by (23.2), may be written as follows:

$$f^*(w) = w - \frac{w(w-1)}{\pi} \iint_D \frac{e^{2i\arg z} t\,\mu_0(|z|)}{z(z-1)(z-w)} dxdy + \mathcal{O}(t^2).$$

Thus in the polar coordinates $|z| = \rho$, $\arg z = \vartheta$ we have

$$f^*(w) = w - \frac{w(w-1)}{\pi} \int_s^{s+\Delta s} \frac{e^{2i\vartheta} t\,\mu_0(\rho)\rho d\rho}{\rho e^{i\vartheta}(\rho e^{i\vartheta}-1)(\rho e^{i\vartheta}-w)} d\rho + \mathcal{O}(t^2)$$

$$= w - t\frac{w(w-1)}{\pi i} \int_s^{s+\Delta s} \frac{\mu_0(\rho)}{\rho} \int_{\partial\Delta\rho} \frac{dz}{(z-1)(z-w)} d\rho + \mathcal{O}(t^2).$$

Hence, by (19.6), we obtain (24.1).

Corollary 25 enables to find extremal functions within any class $\bigcup_R E_Q^{r,R}$ (fixed thereafter) for a relatively wide family of functionals. Let $z_1, \ldots, z_n$ be fixed points satisfying the conditions $r \le |z_n| < |z_{n-1}|, \ldots, |z_2| < z_1 = 1$. For arbitrary mappings $\tilde{f}$ and f of the class in question, let

$$\tilde{f}(z_k) = \tilde{w}_k, \; f(z_k) = w_k, \; k = 1, \ldots, n.$$

Clearly, $\tilde{w}_1 = w_1 = 1$. Consider an arbitrary real-valued function F of the class C^1, defined in a neighbourhood D of $(w_2, \ldots, w_n)$ in $\mathbb{C}^{n-1}$. Suppose that

$$(24.2) \quad A = w_m F_{\tilde{w}_m}(w_2, \ldots, w_n) + w_n F_{\tilde{w}_n}(w_2, \ldots, w_n) \ne 0, \quad m = 2, \ldots, n.$$

Finally, let $\mathcal{F}[\tilde{f}] = F(\tilde{w}_2, \ldots, \tilde{w}_n)$ for $(\tilde{w}_2, \ldots, \tilde{w}_n) \in D$. Then we have (Ławrynowicz [4]):

COROLLARY 26. *If the above assumptions are satisfied and the functional $\mathcal{F}$ attains an extremum for $\tilde{f} = f$, then this extremum is also attained for any function f_1 defined by the formulae $f_1(z) = f(z)$ for $|z_n| < |z| \le 1$ and $f_1(z) = w_n f_0(z/z_n)$ for $|z| \le |z_n|$, where f_0 is an arbitrary mapping of the class in question. Moreover, f is either the identity mapping or*

$$(24.3) \quad f(z) = w_m |z/z_n|^{B_m(z_2, \ldots, z_n; \varepsilon_m)} e^{i\arg(z/z_m)}$$

for $|z_{m+1}| \le |z| \le |z_m|$, $m = 1, \ldots, n-1$,

where

$$(24.4) \quad B_m(z_2, \ldots, z_n; \varepsilon_m) = \tfrac{1}{2}(Q + \tfrac{1}{Q}) - \tfrac{1}{2}\varepsilon_m(Q - \tfrac{1}{Q})\exp[-i\arg\sum_{k=m+1}^{n} w_k F_{\tilde{w}_k}(w_2, \ldots, w_n)],$$

$\varepsilon_m = 1$ <u>or</u> -1, <u>and the branch of</u> $\arg[f(z)/z]$ <u>is chosen for</u> $|z_{m+1}| \leq |z| \leq |z_m|$ <u>so that</u> $f(z) \to w_m$ <u>as</u> $z \to z_m$.

<u>Proof</u>. Let us consider the Gâteaux variation $\check{\tilde{f}} - f + o(t)$ of the function f according to the formula (24.1), where $f^* = \tilde{f} \circ \check{f}$; $\mu = t\mu_0$ being the complex dilatation of f^*, within the class of mappings in question, where s, Δs are fixed and they satisfy the condition $r' < s < 1 - \Delta s$, in such a way that the mapping obtained be conformal outside the annulus $\{s: s < |w| < s + \Delta s\}$. At the moment we do not assume that f is an extremal function. Besides, let us denote the complex dilatation of $\check{f}$ and $\check{\tilde{f}}$ by μ_* and $\tilde{\mu}_*$, respectively, and let $\varkappa = \mu \circ \check{f}^*$, $\varkappa_* = \mu_* \circ \check{f}^*$. We are going to evaluate the effect of this variation for the dilatation μ_*, i.e. to express the difference $|\tilde{\mu}_*| - |\varkappa_*|$

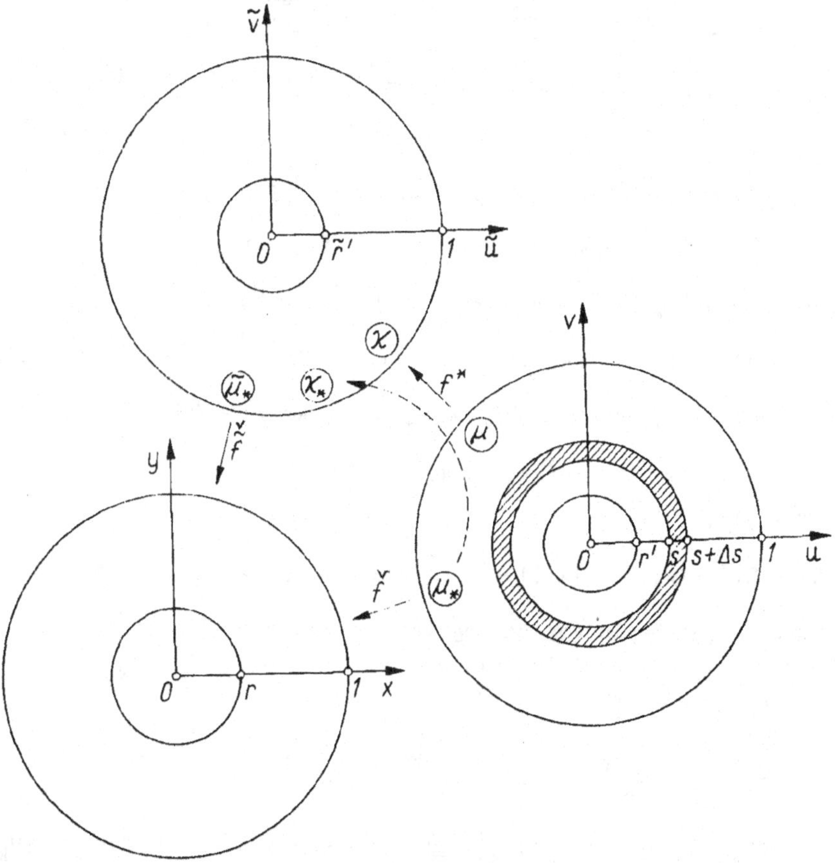

Fig. 4

by $\varkappa$ and $\varkappa_*$ with the exactness $\mathcal{O}(|\varkappa|^2)$.

Outside $\{s: s<|w|<s+\Delta s\}$ there will be not any change of $|\mu_*|$, and in order to calculate the variation inside this annulus we have to express in the differential $\underline{d}\check{f} = \check{f}'_w(dw+\mu_* d\bar{w})$ [we accept the convention: $\underline{d}\check{f}(w) = d\check{f}(w)(dw, d\bar{w})$] $dw = \underline{d}\check{f}(\tilde{w})$ by $d\tilde{w}$, $d\bar{\tilde{w}}$ for $w = \check{f}^*(w)$. To this end we notice that, by (24.1) and $\mu = t\mu_o$,

$$\underline{d}f^* = [1+\mathcal{O}(|\int_s^{s+\Delta s}\mu(\rho)\rho^{-1}d\rho|)](dw+\mu d\bar{w}) \text{ a.e.}$$

Consequently

$$\underline{d}\check{f}^* = (1+\eta_1)d\tilde{w} - (1+\eta_2)\varkappa d\bar{\tilde{w}},\ \eta_1 = \mathcal{O}(|\varkappa|),\ \eta_2 = \mathcal{O}(|\varkappa|).$$

Thus

$$\underline{d}\check{\check{f}} = \underline{d}(\check{f}\circ\check{f}^*) = (\check{f}'_w\circ\check{f}^*)[\underline{d}\check{f}^* + (\mu_*\circ\check{f})\overline{\underline{d}\check{f}^*}]$$

$$= (\check{f}'_w\circ\check{f}^*)[(1+\eta_1)d\tilde{w} - (1+\eta_2)\varkappa d\bar{\tilde{w}} + \varkappa_*(1+\bar{\eta}_1) - \varkappa_*(1+\bar{\eta}_2)\bar{\varkappa}d\tilde{w}]$$

$$= (\check{f}'_w\circ\check{f}^*)[(1+\eta_1-\bar{\varkappa}\varkappa_* - \bar{\varkappa}\varkappa_*\eta_2)d\tilde{w} + (\varkappa_* + \varkappa_*\bar{\eta}_1 - \varkappa - \varkappa\eta_2)d\bar{\tilde{w}}],$$

whence, by $\eta_1 = \mathcal{O}(|\varkappa|)$ and $\eta_2 = \mathcal{O}(|\varkappa|)$, we obtain

$$|\tilde{\mu}_*| = |(\varkappa_* + \varkappa_*\bar{\eta}_1 - \varkappa - \varkappa\eta_2)/(1+\eta_1 - \bar{\varkappa}\varkappa_* - \bar{\varkappa}\varkappa_*\eta_2)|$$

$$= |\varkappa_* + \varkappa_*\bar{\eta}_1 - \varkappa_*\eta_1 - \varkappa - \bar{\varkappa}\varkappa_*^2| + \mathcal{O}(|\varkappa|^2).$$

Next, in view of the identity $|z| \equiv (\mathrm{re}^2 z + \mathrm{im}^2 z)^{\frac{1}{2}}$ we have, for $f(z) \not\equiv z$,

$$|\tilde{\mu}_*| = |\varkappa_*|[1+2\,\mathrm{re}(\bar{\varkappa}\varkappa_* - \varkappa/\varkappa_*) + \mathcal{O}(|\varkappa|^2)]^{\frac{1}{2}} + \mathcal{O}(|\varkappa|^2)$$

$$= |\varkappa_*|[1+\mathrm{re}(\bar{\varkappa}\varkappa_* - \varkappa/\varkappa_*)] + \mathcal{O}(|\varkappa|^2)$$

and, finally,

$$(24.5)\quad |\tilde{\mu}_*| - |\varkappa_*| = -|\varkappa|(1-|\varkappa_*|^2)\cos(\arg\varkappa_* - \arg\varkappa) + \mathcal{O}(|\varkappa|^2).$$

Now we may begin evaluating the variation of $\mathcal{F}$. Since F is a real-valued function of the class $C^1(D)$, we have $F_{\bar{\tilde{w}}_k} = \overline{F_{w_k}}$ for $k=2, \ldots, n$ and, consequently, by (24.1),

$$\underline{d}F = \sum_{k=2}^{n}(F_{\tilde{w}_k}d\tilde{w}_k + F_{\bar{\tilde{w}}_k}d\bar{\tilde{w}}_k),\ d\tilde{w}_k = \underline{d}\tilde{f}_k(s),\ \tilde{f}_k = \tilde{f}(z_k) = f^*(w_k).$$

$$\underline{d}\tilde{f}_k(s) = 0 \text{ for } s \le |w_k|,\ \underline{d}\tilde{f}_k(s) = -2w_k\mu(s)s^{-1}ds \text{ for } s > |w_k|$$

a.e. in $\{s: r' \le |s| \le 1\}$, where also $\underline{d}F(w_2,\ldots,w_n) = t\partial\mathcal{F}[f]$. Therefore the variation $dF(s)$ can be expressed a.e. in $D_m = \{w: |w_{m+1}| \le |w| \le |w_m|\}$, $m = 1, \ldots, n-1$, by the formula

$$(24.6)\quad \underline{d}F(w_2,\ldots,w_n)(s) = -2\sum_{k=m+1}^{n} w_k F_{\tilde{w}_k}(w_2,\ldots,w_k)\mu(s)s^{-1}ds,\quad s = |w|,$$

and hence, by (24.2), we get

$$(24.7)\quad \underline{d}F(w_2,\ldots,w_n)(s) = -2|A||\mu(s)|\cos[\arg\mu(s) + \arg A]s^{-1}ds.$$

Suppose that $|\mu_*| < (Q-1)/(Q+1)$ on a set K of positive measure. Since this set is the union of K_ε, $\varepsilon > 0$, where $|\mu_*| \le (Q-1)/(Q+1) - \varepsilon$, then there is a number $\varepsilon^* > 0$, for which K_{ε^*} is of positive measure. Consequently, comparing the relations (24.5) and (24.6) we see that, in the case where $|\mu_*| < (Q-1)/(Q+1)$, it is admissible that the variation (24.7) is taken with an arbitrarily small measurable function $|\mu|$ and that the value $\underline{d}F(w_1,\ldots,w_n)(s)$ may have an arbitrary sign. Hence, if f is an extremal function, the dilatation μ_* of $\check{f}$ cannot satisfy the condition $|\mu_*| < (Q-1)/(Q+1)$ on a set of positive measure. Since we have $|\mu_*| \le (Q-1)/(Q+1)$ a.e., then also a.e. we have

$$(24.8)\quad |\mu_*(w)| = (Q-1)/(Q+1)$$

and, moreover, by (24.5), (24.7), and the definition of $\varkappa$ and $\varkappa_*$,

$$(24.9)\quad \arg\mu_*(w) = -\arg\sum_{k=m+1}^{n} w_k F_{\tilde{w}_k}(w_2,\ldots,w_k) + \tfrac{1}{2}\pi(1-\varepsilon_m)$$

a.e. in D_m, $m = 1, \ldots, n-1$, where $\varepsilon_m = 1$ or -1.

In turn, since – by (24.8) and (24.9) – the dilatation μ_* is constant a.e. in D_m, $m = 1, \ldots, n-1$, we denote it by γ_m there. Hence, in view of Lemma 17 (especially condition (v) which – if applied to $\check{f}$ – gives

$$(24.10)\quad \check{f}(w) = \exp\Big(-\int_{|w|}^{1} \frac{1+\mu_*(s)}{1-\mu_*(s)}\,\frac{ds}{s} + i\arg w\Big),\quad r' \le |w| \le 1),$$

we get

$$\check{f}(w) = z_m|w/w_m|^{(1+\gamma_m)/(1-\gamma_m)}e^{i\arg(w/w_m)},\quad |w_{m+1}| \le |w| \le |w_m|,$$

where the branch of $\arg[\check{f}(w)/w]$ is chosen so that $\check{f}(w) \to z_m$ for $w \to w_m$. Finally we notice that if for an arbitrary function of the form (24.10) we set $\check{R}(w) = |\check{f}(|w|)|$ and $-\Theta(|w|) = \arg\check{f}(|w|)$, then $|f(z)| = R(|z|)$ and $\arg[f(z)/z] = \Theta \circ R(|z|)$, where $\arg[f(z)/z] =$

$-\arg[z/f(z)]$. Hence the formulae (24.3) follow immediately, where

$$\beta_m(z_2,\ldots,z_n;\varepsilon_m) = \left(1 - i\,\mathrm{im}\frac{1+\gamma_m}{1-\gamma_m}\right)/\mathrm{re}\frac{1+\gamma_m}{1-\gamma_m}.$$

Therefore, by the conditions (24.8), (24.9), and $\mu_*(w) = \gamma_m$ which hold a.e. in D_m, $m = 1, \ldots, n-1$, respectively, we obtain the relations (24.4), but we cannot decide a priori which system of ε_m, $m = 1, \ldots, n-1$, corresponds to the case of the minimum, and which to the case of the maximum of the given functional F.

25. Schiffer's variational method

Let

$$(25.1)\quad h(w; w_0, s) = \begin{cases} s^2(w - w_0)^{-1} & \text{for } |w - w_0| \geq s, \\ (\bar{w} - \bar{w}_0)(2 - s^{-2}|w - w_0|^2) & \text{for } |w - w_0| < s. \end{cases}$$

It is easily seen that $h(\ ; w_0, s)$ is of the class C^1 on $\mathbb{E}$. Let Γ_0 be a fixed Jordan curve and let $(w_1, \ldots, w_n)$ be an arbitrary system of points situated inside Γ_0. Suppose that the discs $K_j = \{w: |w - w_j| \leq s_j\}$, $j = 1, \ldots, n$, are disjoint and lie inside Γ_0. For $t > 0$ take n complex numbers b_j such that $|b_j| < t$ and form the variational function

$$(25.2)\quad g(w,\bar{w}) = w + \sum_{j=1}^{n} b_j h(w; w_j, s_j).$$

If t is sufficiently small, g is univalent on Γ_0 and its Jacobian positive inside Γ_0. Hence (cf. Goursat [1], vol. I, pp. 388-389 and the proof of Lemma 13) g is a C^1-homeomorphism on $\mathbb{E}$ and $g[\mathbb{E},\mathbb{E}] = \mathbb{E}$.

Following Schiffer [2], we introduce, for the sake of convenience in calculations, the dilatation $R_f(z)$ defined by the formula

$$(25.3)\quad R_f(z) = [|f_z(z)|^2 + |f_{\bar{z}}(z)|^2]/[|f_z(z)|^2 - |f_{\bar{z}}(z)|^2]$$

at any point z, at which there exist the derivatives $f_z(z)$ and $f_{\bar{z}}(z)$. It is related to the dilatation

$$p_f(z) = [|f_z(z)| + |f_{\bar{z}}(z)|]/[|f_z(z)| - |f_{\bar{z}}(z)|]$$

by a simple formula

$$R_f(z) = \tfrac{1}{2}[p_f(z) + 1/p_f(z)].$$

If f is a diffeomorphism defined in a domain D and $R_f(z) \leq M$ in D, then f is Q-qc and $Q = M + (M^2 - 1)^{\frac{1}{2}}$. Conversely, the boundedness of p_f yields the boundedness of R_f.

Suppose that f is a qc mapping of D. Hence the composed mapping $f^* = g \circ f$ or, more explicitly,

$$(25.4) \quad f^*(z,\bar{z}) = g(f(z,\bar{z}), \overline{f(z,\bar{z})})$$

has at the differentiability points of f the dilatation R_{f^*} given (for z such that $f(z,\bar{z}) \in K_j$) by

$$(25.5) \quad R_{f^*}(z) = R_f(z) + 8\left(1 - \frac{|f(z) - w_j|^2}{s_j^2}\right) \frac{\operatorname{re}[\bar{b}_j f_z(z) f_{\bar{z}}(z)]}{|f_z(z)|^2 - |f_{\bar{z}}(z)|^2} + \mathcal{O}(t^2).$$

Therefore, if $t > 0$ is sufficiently small, then the dilatation increases arbitrarily little, so the mapping f^* defined by (25.4) and (25.2) can be used as a comparison function. However, as it is remarked by Renelt [1, 3], if $n \to +\infty$, also R_{f^*} may tend to $+\infty$. In order to push this complication aside, it is sometimes more convenient to utilize a variational formula given in the form of an integral. Namely, we have

<u>THEOREM 25</u>. <u>Suppose that $\mu_0 : \mathbb{C} \to \mathbb{C}$ is a measurable function with compact support and $\|\mu_0\|_\infty < 1$. Let $w \mapsto f^*(w,t)$ be the unique qc self-mapping of $\mathbb{C}$ generated by the complex dilatation $\mu = t\mu_0$, $0 < t \le 1$, and subject to the normalization $f^*(w,t) - w \to 0$ as $w \to \infty$. Then</u>

$$(25.6) \quad f^*(w,t) = w - (t/\pi) \iint_{\mathbb{C}} \mu_0(z)(z - w)^{-1} dxdy + \mathcal{O}(t^2),$$

<u>where $\mathcal{O}(t^2)/t^2$ is uniformly bounded in $\mathbb{C}$ as $t \to 0+$.</u>

<u>Remark 4</u>. Theorem 25 is due in this form to Lehto who gave in [6] a very simple proof of it. In the particular case where

$$\mu_0(w) = b_j \quad \text{for } w \in K_j, \ j = 1, \ldots, n; \quad \operatorname{spt} \mu_0 = \bigcup_{j=1}^{n} K_j,$$

K_j being disjoint closed discs and b_j being complex constants with $|b_j| < 1$, $j = 1, \ldots, n$, this theorem was proved by Renelt [1, 3].

<u>Proof of Theorem 25</u>. Let $\mu_0 : \mathbb{C} \to \mathbb{C}$ be a measurable function with $\operatorname{spt} \mu_0 \subset E$, where E is compact, and $\|\mu_0\|_\infty < 1$. Let f be the qc mapping generated by μ_0 and subject to the normalization $f(z) - z \to 0$ as $z \to \infty$. Then, by Lemma 4,

$$(25.7) \quad f(z) = z + Tg(z),$$

where $g = f_{\bar{z}}$ can be found as in Lemma 10:

$$(25.8)\quad g = \sum_{n=1}^{+\infty} h_n,\ h_1 = \mu,\ h_n = \mu S h_{n-1} \text{ for } n = 2, 3, \ldots,$$

the transformations S and T being defined as in those lemmas. The series in (25.8) is absolutely convergent in L^p as soon as $p > 2$ is chosen so that $\|\mu_0\|_\infty \|S\|_p < 1$; cf. Lemma 10. Moreover,

$$(25.9)\quad \|h_n\|_p \leq C(\|\mu_0\|_\infty \|S\|_p)^n$$

which easily follows by the Calderón-Zygmund inequality (6.1) in Lemma 8. If h has bounded support contained in E and h is in L^p, then

$$(25.10)\quad |Th(z)| \leq C_1 \|h\|_p,$$

where C_1 depends on p and E only.

From (25.7)-(25.10) we obtain

$$(25.11)\quad f(z) = z + T\mu(z) + \sum_{n=2}^{+\infty} T h_n(z).$$

Combining (25.9) and (25.10) we obtain

$$(25.12)\quad \sum_{n=2}^{+\infty} |T h_n(z)| \leq C_2 \sum_{n=2}^{+\infty} (\|\mu_0\|_\infty \|S\|_p)^n.$$

This shows that the series in (25.11) is uniformly convergent in $\mathbb{C}$. Let us now replace μ_0 by $t\mu_0$, $0 \leq t \leq 1$, and let $f^*(\ ,t)$ be the mapping generated by $t\mu_0$. Then it follows from (25.12) that

$$\sum_{n=2}^{+\infty} |T h_n(w,t)| = \mathcal{O}(t^2) \text{ as } t \to 0+,$$

the bound being uniform in w, and finally we arrive at (25.6), where $\mathcal{O}(t^2)/t^2$, $0 < t \leq 1$, is uniformly bounded in $\mathbb{C}$ as $t \to 0+$. This ends the proof.

In the case where f is a differentiable mapping, the composed mapping $f^* = g \circ f$ is, as it can easily be verified, differentiable as well and its dilatation R_{f^*} may be expressed by the formula

$$(25.13)\quad R_{f^*}(z) = R_f(z) + \frac{4t\,\mathrm{re}[\bar{b}_1 f_z(z) f_{\bar{z}}(z)]}{|f_z(z)|^2 - |f_{\bar{z}}(z)|^2} + \mathcal{O}(t^2)$$

for $z \in \check{f}[K_j]$ (cf. Remark 4). Formula (25.13) involves the points of discontinuity of f^*_z and $f^*_{\bar z}$ on $\operatorname{fr} K_j$, but it is more convenient than (25.5) in the sense that if the number n of discs K_j tends to infinity, the dilatation R_{f*} exceeds R_f arbitrarily little for $|b_j| < t$ and t sufficiently small.

It appears that Schiffer's variational formulae (25.2) and (25.4) can give important information on the extremal functions even if we take $n = 1$. Suppose that a real-valued functional F, differentiable in the sense of Gâteaux, attains an extremum for a qc mapping of the class C^1 within the family of mappings $\tilde f$ which are qc in a domain D with $R_{\tilde f}(z) \le M$ for $z \in D$, M being a finite constant. Let

$$(25.14)\quad g(w) = w + b_0 h(w; w_0, s),$$

where h is defined by (25.1). Obviously, it is a particular case of (25.2). We are going to prove that $R_f(z) = M$ for $z \in D$.

In fact, let z D and let $R_f(z_0) = M - c$, $c > 0$. We choose a disc $K_0 = \Delta^s(w_0)$, $w_0 = f(z_0)$, and the function $h(\ ; w_0, s)$ in (25.1) so that the dilatation $R_{f*}(z)$ of the composed function $f^* = g \circ f$ does not exceed $M - \frac{1}{2}c$ for $z \in \check{f}[K_0]$; this can be achieved in view of the assumption that f is of the class C^1 in D. The mapping g is conformal outside K_0, so f^* belongs indeed to the class in question. Hence, by (23.1), the comparison function f^* satisfies

$$(25.15)\quad F[f^*] = F[f] + \operatorname{re}\{b_0 \partial F[f][h(f; w_0, s)]\} + \mathcal{O}(|b_0|^2).$$

Therefore, since b_0 can be arbitrarily chosen, we obtain

$$(25.16)\quad \partial F[f][h(f; w_0, s)] = 0_D$$

for every $w_0 = f(z_0)$ such that $R_f(z_0) < M$.

Further discussion relies upon the form of the first Gâteaux variation. If we suppose it to be continuous in the topology of almost uniform convergence in D, then, by (25.1) and the linearity of $\partial F[f]$, we obtain

$$(25.17)\quad \partial F[f][h(f; w_0, s)] = \partial F[f][s^2/(f - w_0)] = s^2 \partial F[f][1/(f - w_0)].$$

If we assume that the function A, defined in its natural domain of definition by

$$(25.18)\quad A(\omega) = \partial F[f][1/(f - \omega)]$$

is meromorphic and not identically zero, then (25.17) contradicts

(25.16). Relation (25.16) will be satisfied not only at w_0, but, in view of the continuity of f, also in a disc centred at w_0. Since a meromorphic function vanishing in a disc is identically zero in the whole domain of definition, then we get a contradiction with the hypothesis that A is not identically zero. This contradiction, obtained under the hypothesis $R_f(z_0) < M$, implies $R_f(z) = M$ identically for any function that maximizes F.

The above argument may be repeated almost without changes in the case where f is of the class C^1 in $D \setminus H$, where H is an exceptional set being the finite union of one-point sets and Jordan arcs. Thus we obtain

THEOREM 26. Suppose that a real-valued functional F, differentiable in the sense of Gâteaux, attains an extremum for a mapping f within the class of qc mappings of a domain D with dilatation R_f uniformly bounded by a constant M, and f is of the class C^1 in D [correspondingly: in $D \setminus H$, H being as above]. Then, if the first Gâteaux variation of F satisfies (25.17) and (25.18), where A is meromorphic in its natural domain of definition, then the dilatation of R_f is constant in D [correspondingly: in $D \setminus H$].

Different variants of this theorem appear in several earlier papers, for instance in Teichmüller [1]. Formula (25.5), eventually (25.13), implies that if $\mathrm{re}(\bar{b}_0 f_z f_{\bar{z}})$ is negative at a point z_0 at which the extremal mapping f possesses the partial derivatives f_z and $f_{\bar{z}}$, then, even in the case where $R_f(z_0) = M$, we can form a comparison function f^* of the form $g \circ f$ with g given by (25.14). The function f^* has in the whole, sufficiently small neighbourhood of z_0 the dilatation R_f which does not exceed M and, moreover, outside of this neighbourhood, in view of the conformality of g, we have $R_f = R_{f^*}$. Taking into account (25.17), we conclude that $(A \circ f) f_z f_{\bar{z}} \geq 0_D$. Since, by Theorem 26, $|f_{\bar{z}}/f_z| = [(M-1)/(M+1)]^{\frac{1}{2}} = \mathrm{const}$, we finally obtain

$$(25.19) \qquad (A \circ f)^{\frac{1}{2}} f_{\bar{z}} = [(M-1)/(M+1)]^{\frac{1}{2}}\, \overline{(A \circ f)^{\frac{1}{2}} f_z}\,.$$

Thus we have proved the following

COROLLARY 27. If the assumptions of Theorem 26 are fulfilled and A is holomorphic, then the extremal function f satisfies the differential equation (25.19).

If we set $\Psi(\omega) = \int [A(\omega)]^{\frac{1}{2}} d\omega = \int^{\omega} [A(w)]^{\frac{1}{2}} dw$ and $C(z,\bar{z}) = \Psi \circ f(z)$, then the equation (25.19) can be rearranged to the form

$$(\partial/\partial\bar{z})[(M+1)^{\frac{1}{2}}C(z,\bar{z}) - (M-1)^{\frac{1}{2}}\overline{C(z,\bar{z})}] = 0.$$

By Lemma 4 this implies that the function $2\Phi = (M+1)^{\frac{1}{2}}C - (M-1)^{\frac{1}{2}}\bar{C}$ is holomorphic in D. Therefore the extremal function f satisfies in D the functional equation

$$(25.20) \quad \Psi \circ f = \Phi + [(M-1)/(M+1)]^{\frac{1}{2}}\bar{\Phi}, \quad \Psi(\omega) = \int [A(\omega)]^{\frac{1}{2}}d\omega,$$

where A is the holomorphic function related to the first Gâteaux variation of $\mathcal{F}$ by the formula (25.18).

Equation (25.20) has the form $\Psi \circ f = L \circ \Phi$, where L is an affine mapping. By virtue of the local invertibility of mappings realized by holomorphic functions, we conclude from (25.20) that, locally in D, apart from an isolated set, we have

$$(25.21) \quad f = \check{\Psi} \circ L \circ \Phi.$$

Thus the function f, extremal for $\mathcal{F}$, is locally, apart from an isolated set, the composition of two conformal mappings and of an affine mapping. Evidently the dilatation R_L^* is constant in D.

The fundamental role of qc mappings of the form (25.1) in extremal problems of the theory of qc mappings was first discovered by Grötzsch [1, 2] and then intensively studied by Teichmüller [1]. As an example of mappings of this form may serve Teichmüller mappings considered in Theorems 13-16 which, by those theorems, satisfy the relation (17.8), i.e. (25.20) with $L \circ \Phi = \Phi + t\bar{\Phi}$ or, equivalently, $\check{L} \circ \Psi = (1-t^2)^{-1}(\Psi - t\bar{\Psi})$. The function $\Psi = \int [A(\omega)]^{\frac{1}{2}}d\omega$ corresponds to the quadratic differential $[A(\omega)]^{\frac{1}{2}}d\omega$ in D, so Corollary 27 points out at the same time the relationship between extremal mappings and quadratic differentials. Besides, this correspondence indicates further possibilities of generalization: it is not essential for $A(\omega)$ to be single-valued and then the results can be extended to quadratic differentials on Riemann surfaces (cf. the next section). Also these facts were first discovered by Teichmüller in his fundamental paper [1].

It is clear that Corollary 27 is incomplete in the sense that we do not know whether the extremal mapping in question is of the class C^1 indeed. Nevertheless the method can be applied almost without changes in the case where the extremal mapping is of the class C^1 in $D \setminus H$ with H as before. In some particular cases it has been already proved that the extremal function has such a form indeed (cf. e.g. Renelt [1, 3]). Then the dilatation R_f is constant in $D \setminus H$.

III. A review of variational methods and basic applications

The method of Schiffer is extensively described in the book by Schober [1], and the particularly important contribution of Renelt - in his detailed paper [3] being in principle only a more easily accessible version of [1]; cf. also Schiffer and Schober [3, 4, 6], and Renelt [4].

26. Extremal quasiconformal mappings

Teichmüller's theorem and applications

Suppose that $\mathcal{T}$ is a nonempty class of homeomorphisms $\tilde{f}: W \to W'$, where W and W' are fixed Riemann surfaces. If $\mathcal{T}$ contains mappings $\tilde{f}$ of bounded maximal dilatation $Q[f]$, we may consider in $\mathcal{T}$ the problem of finding a mapping f such that $Q[f] = \inf_{\tilde{f} \in \mathcal{T}} Q[\tilde{f}]$, i.e. a so-called <u>extremal</u> <u>qc</u> <u>mapping</u>.

A problem of this type was first investigated by Grötzsch [1] for homeomorphisms of a closed rectangle R onto a rectangle R', carrying the vertices of R into the vertices of R'. Then the extremal qc mapping was shown to be an affine mapping.

Systematic investigations of extremal qc mappings were initiated by Teichmüller in the paper [1] which played a similarly inspiring role in the theory of conformal and qc mappings as the paper of O. Hahn and F. Strassmann, published in the same volume, played in the nuclear physics. In that paper Teichmüller characterized the general form of extremal qc mappings and he first discovered their relationship with quadratic differentials.

Let us recall that under a <u>quadratic</u> <u>differential</u> on W we mean an assignment to an arbitrary local parameter z for W of a meromorphic function φ such that if some local parameters z and z_1 describe a neighbourhood of a point P on W, then the corresponding functions φ and φ_1 satisfy the formal equality $\varphi_1 dz_1^2 = \varphi\, dz^2$, i.e. $z_1 = g(z)$ and $(\varphi_1 \circ g)(g')^2 = \varphi$. Hence with φ also $\lambda\varphi$, $\lambda = \text{const}$, is a quadratic differential.

Teichmüller [1] did not give binding proofs of his results, basing rather on examples and heuristic considerations. A complete proof of his fundamental theorem characterizing extremal qc mappings of closed Riemann surfaces onto itself was given only in his paper [3]. Other proofs were later given by Ahlfors [2], Bers [2], and Kruškal [3]. For a beginner we would recommend a proof given by Kruškal in his book [8], pp. 46-60. Teichmüller's theorem is a certain general variational principle in the theory of qc mappings and is directly related to Corollary 27. In order to formulate Teichmüller's theorem we need some preliminaries.

It is well known that any orientable closed surface which is finite (i.e. has a finite triangulation) is homeomorphic to a sphere endowed with some g handles (the so-called <u>normal form of Klein</u> of the surface). The number g is called the <u>genus</u> of the surface in question. Thus we may speak about the genus of a compact Riemann surface since its support is homeomorphic to an orientable closed surface which is finite.

The general UNIFORMIZATION THEOREM proved by F. Klein, H. Poincaré and P. Koebe (cf. e.g. Springer [1], p. 226) says that <u>for an arbitrary Riemann surface</u> W <u>there exist a canonical domain</u> G <u>(which is the closed plane</u> $\mathbb{E}$, <u>open plane</u> $\mathbb{C}$ <u>or open unit disc</u> $\operatorname{int}\Delta$ <u>) and a single-valued analytic function (in other words: a holomorphic function)</u> F: $G \to W$ <u>with the following properties</u>:

a) F <u>is a local homeomorphism</u>;

b) <u>to every point</u> $P \in W$ <u>there corresponds a sequence</u> (z_n) <u>of points in</u> G <u>(the so-called sequence of equivalent points) such that</u> $F(z_n) = P$, $n = 1, 2, \ldots$;

c) <u>there is a group</u> Γ <u>of linear fractional transformations of</u> G <u>onto itself, isomorphic to the fundamental group of</u> W <u>(i.e. the group of homotopy classes of closed curves on</u> W<u>) such that every linear fractional transformation</u> $S \in \Gamma$ <u>carries an arbitrary point of</u> G <u>into an equivalent point</u>;

d) <u>apart from the identity</u> id_G <u>no transformations</u> $S \in \Gamma$ <u>have invariant points in</u> G.

In the canonical domain G we may distinguish a maximal connected set G^* which contains no pairs of equivalent points. This is the so-called <u>fundamental domain</u> of the Riemann surface W. In the case $G = \mathbb{E}$ which corresponds to $g = 0$, Γ contains exactly one element id_G. Then $G = G^*$. In the case $G = \mathbb{C}$ we have to distinguish three cases:

(i) W is conformally equivalent to a sphere without the point ∞, Γ contains id_G only, and $G = G^*$;

(ii) W is conformally equivalent to a cylinder, G^* is a strip $\{z\colon 0 < \operatorname{re} z \leq b\}$, and Γ is the group of translations $z' = z + nb$, n being an integer;

(iii) W is a surface of genus 1, i.e. W is conformally equivalent to a torus, G^* is a parallelogram, and Γ is the group of translations $z' = z + ma + nb$, where $a, b \in \mathbb{C}$ and $\operatorname{im}(a/b) \neq 0$, m and n being integers.

To every closed surface of genus $g > 1$ there corresponds the canonical domain $G = \text{int}\,\Delta$ and the group Γ is then the so-called Fuchsian group, i.e. a countable group of non-euclidean motions S_*: $\text{int}\,\Delta \longrightarrow \text{int}\,\Delta$ which have no invariant points in $\text{int}\,\Delta$ for $S_* \neq \text{id}_G$. This group is determined with the exactness to conjugation in the group of all conformal automorphisms of Δ. The group Γ will be called (for $g > 1$) the Fuchsian group of the Riemann surface W.

The groups Γ of W and Γ' of W' are identical iff W and W' are conformally equivalent. A variable $z \in G$ may be regarded as a local parameter of W in a sufficiently small neighbourhood of the point $P = F(z)$. If $\tilde{g}$ is a (multivalued) analytic function on W, continuable along any curve of W, the monodromy principle implies that $\tilde{g} \circ F$ is single-valued in G. Thus the mapping F "uniformizes" any function $\tilde{g}$ which is analytic on W in the sense that $\tilde{g} \circ F$ is holomorphic.

Let us confine ourselves to Riemann surfaces of genus $g > 1$. Suppose that W' is a Riemann surface whose support is homeomorphic to the support of W. We are going to introduce now the notion of homotopic homeomorphisms f_k: $W \longrightarrow W'$, $k = 1, 2$. The mappings f_1 and f_2 are said to be homotopic if there is a conformal mapping h: $W' \longrightarrow W'$ such that $\check{f}_2 \circ h \circ f_1$: $W \longrightarrow W$ is homotopic to id_W. Denote by $\mathsf{T}(\alpha)$ a fixed class of homotopic homeomorphisms of W onto W'. Let $\tilde{\omega} \in \text{int}\,\Delta$ be a local parameter of W' determined in analogy to the parameter z, determined before for W. Any mapping $f \in \mathsf{T}(\alpha)$ generates a mapping $\tilde{\omega}$: $\text{int}\,\Delta \longrightarrow \text{int}\,\Delta$ which transforms the fundamental domain G^* of W into $G^{*\prime}$ of W' and, moreover, satisfies the identity

$$\tilde{\omega}(S_* z) = S^\alpha \tilde{\omega}(z),$$

where $S_* \in \Gamma$ and S^α is an element of the Fuchsian group Γ' of W', uniquely determined by S_* and $\mathsf{T}(\alpha)$. If, in particular, f is qc (we write then $f \in \mathsf{Q}(\alpha)$), then $\tilde{\omega}$ is qc as well. It can be proved that the class $\mathsf{Q}(\alpha)$ is nonempty and that it contains an extremal qc mapping ω with a minimal maximal dilatation. With the above notation TEICHMÜLLER's THEOREM may be formulated as follows (cf. Ahlfors [2]):

Suppose that W and W' are compact Riemann surfaces of genus $g > 1$ and ω is an extremal qc mapping corresponding to $\mathsf{T}(\alpha)$. Then either ω is a holomorphic function or there exist a quadratic differential $\varphi\, dz^2$ on W and a constant t, $0 < t < 1$, such that at any point $z \in \text{int}\,\Delta$ where $\varphi(z) \neq 0$ the function ω is differentiable, has nonvanishing derivatives ω_z, $\omega_{\bar{z}}$, and

(26.1) $\omega_{\bar{z}}/\omega_z = t\bar{\varphi}/|\varphi|$.

26. Extremal quasiconformal mappings. Teichmüller's theorem

Let us notice that taking in the quadratic differential $\varphi\, dz^2$, z and $z_1 = S_* \circ z$ as local parameters, according to the definition we obtain the identity $\varphi \circ S_* = \varphi S_*'^{-2}$. The second remark is that, in addition to the motivation given in the preceding section, Teichmüller's theorem gives another motivation for considering Teichmüller mappings. Of course there is a deep connection between these motivations, partially enlightened by Theorems 12-15.

Now we procced to give a third motivation. Any qc mapping of the unit disc $\operatorname{int} \Delta$ onto itself can be, by Corollary 2 and Theorem 4, homeomorphically continued onto its closure Δ. Hence Teichmüller's theorem naturally suggests the problem of finding an extremal qc mapping within <u>the class $\mathfrak{Q}_f$ of qc mappings of Δ onto itself such that their restrictions to</u> $\operatorname{fr} \Delta$ <u>are identical with</u> $f|\operatorname{fr} \Delta$, f <u>being a given qc mapping</u> (and within the corresponding classes $\mathfrak{Q}_f^{r,R}$).

Let $\mathcal{T}$ denote the class of all sense-preserving homeomorphisms $f\colon \Delta \to \Delta$ having locally in $\operatorname{int} \Delta$, except perhaps for an isolated set H, the form (25.21), where Φ and Ψ are conformal mappings and L is an affine mapping. In other words we suppose that every point of $\operatorname{int} \Delta \setminus H$ has a neighbourhood U with a property that there exist conformal mappings Φ and Ψ of the sets U and $V = f[U]$, and an affine mapping L such that (25.21) holds with $f|U$ substituted for f. From this definition it follows that the dilatation of L, and thus also of f, is constant in $\operatorname{int} \Delta \setminus H$. Suppose that L is not linear.

It can be shown (cf. Strebel [2], Part II) that the function Φ can be continued along any curve avoiding the set H in $\operatorname{int} \Delta$. Continuation of an element Φ_1 along any closed curve leads to an element $\Phi_2 = a\Phi_1 + b$, where $a \neq 0$ and $\operatorname{im} a = 0$. Thus under this continuation with an element Φ_1' we may associate the elemant $\Phi_2' = a\Phi_1'$. If we suppose that for the continuation along any closed curve $a = 1$ or -1, then $\Phi_2'^2 = \varphi$ becomes a function holomorphic in $\operatorname{int} \Delta$ outside H. Then the complex dilatation of f will have a.e. the form $t\bar{\varphi}/|\varphi|$, where $0 \leq t < 1$ and φ is a function meromorphic in $\operatorname{int} \Delta$ whose only singularities may be points of H. If we suppose that

$$(26.2) \qquad \|\varphi\|_1 = \iint_{\Delta} |\varphi|\, dxdy < +\infty,$$

then the above described singularities will be at most poles of the first order. In this way we have arrived exactly at Teichmüller mappings according to the definition given in Section 17.

Strebel [2] proved that <u>if</u> $f\colon \Delta \to \Delta$ <u>is a Teichmüller mapping</u>

with φ <u>holomorphic in</u> $\operatorname{int}\Delta$ <u>and satisfying</u> (26.2), <u>then</u> f <u>is the only extremal qc mapping</u> (<u>i.e.</u> <u>a qc mapping of the least maximal dilatation</u>) <u>within the class</u> Q_f. Strebel's result is in a way the best possible, since in Part I of the same paper he gave examples of Teichmüller mappings being not extremal or being extremal but not the only extremal mappings. In both cases the condition (26.2) is not satisfied, i.e. $\|\varphi\|_1 = +\infty$.

The problem of characterizing complex dilatations for extremal qc mappings within the class Q_f has also been investigated by R. S. Hamilton [1], and by Reich and Strebel [3]. In particular Hamilton obtained a necessary condition for a measurable function μ to be the complex dilatation of an extremal qc mapping within Q_f. Let

$$(26.3)\qquad H_*[\mu] = \sup\left|\iint_\Delta \mu\,\varphi\;dxdy\right|,$$

where the supremum is taken w.r.t. all functions holomorphic in $\operatorname{int}\Delta$, of the norm $\|\varphi\|_1 \le 1$. In the paper quoted Hamilton proved that <u>if</u> g <u>is an extremal qc mapping within the class</u> Q_f, <u>then its complex dilatation</u> μ <u>satisfies the condition</u>

$$(26.4)\qquad \|\mu\|_\infty = H_*[\mu].$$

Under more restrictive assumptions the condition (26.4) was derived by Kruškal [4].

Reich and Strebel [3] proved that the condition (26.4) is also sufficient (a generalization to the case of arbitrary open Riemann surfaces with hyperbolic covering has been obtained recently by Strebel [8] and Šeretov [1]). Namely, their theorem states that <u>if a qc mapping</u> g <u>of the class</u> Q_f <u>satisfies the condition</u> (26.4), <u>then it is an extremal mapping within</u> Q_f. The method of the proof of this theorem is the following. Let E_n denote the disc Δ with distinguished n boundary points $\exp i\theta_j$. Then there exists a uniquely determined extremal qc mapping f_n of the "polygon" E_n onto the "polygon" E_n' with vertices $f(\exp i\theta_j)$ which is a <u>Teichmüller</u> mapping whose complex dilatation has a.e. the form $t_n\overline{\varphi_n}/\varphi_n$, where $0 \le t_n < 1$ and φ_n is a function holomorphic in $\operatorname{int}\Delta$ with a property that along the "sides" of E_n the expression $\varphi_n(z)dz^2$ is real. In addition φ_n has at the points $\exp i\theta_j$ at most poles of the first order and satisfies the condition $\|\varphi_n\|_1 = 1$. Next one has to prove that the sequence (φ_n) tends to the right-hand side of (26.3) under the hypothesis that the lengths of sides of E_n tend to zero. Finally one has to prove that the corresponding sequence (f_n) tends to a qc mapping which is extremal within Q_f. For further results cf. Reich [3-7], Strebel [7], Shibata [1], Šeretov [2, 3], and Bers [5].

As a special case of Q_f let us consider the class Q_{id} of qc mappings of Δ onto itself with invariant boundary points. This class, e.g. with the restriction to Q-qc mappings, is very important and natural from the viewpoint of variational methods. We proceed to explain this with help of an example.

Example 10. Let $f: D \to D'$ be a Q-qc mapping. Suppose that Δ' is a closed disc contained in D' and that the maximal dilatation of f in $\check{f}[\Delta']$ does not exceed $Q - t$, where $0 \le t < 1$. Next we construct a Q^*-qc mapping $g: \Delta' \to \Delta'$ such that $g|\mathrm{fr}\,\Delta' = \mathrm{id}_{\mathrm{fr}\,\Delta'}$ and assume that $Q^* \le Q/(Q-t)$. Let us consider now the mapping $f^*: D \to D'$ defined as follows:

$$f^*(z) = \begin{cases} f(z) & \text{for } z \in D \setminus \check{f}[\Delta'], \\ g \circ f(z) & \text{for } z \in \check{f}[\Delta']. \end{cases}$$

It may be easily observed (cf. Lehto and Virtanen [1], p. 47, or [2], p. 45, Theorem 8.3) that f^* is a Q-qc mapping which may serve as a comparison function in a variational approach. Clearly the fact that, in general, $\Delta' \neq \Delta$ is not essential at all.

Qc mappings of the class Q_{id} were first investigated by Teichmüller [2], as we have already remarked. He found in this class for mappings carrying the origin $z = 0$ into a point d an extremal mapping of the least maximal dilatation. Basing on this result it is possible to derive the exact value of $\sup|f(0)|$ within the subclass of Q_{id} consisting of Q-qc mappings (cf. Krzyż [1]). The relationship between the class Q_{id} and the class of Teichmüller mappings has been investigated in detail in Section 17.

Of course variational methods are applicable as well to the classes of conformal mappings with qc extension considered in Sections 21 and 22. The results of a special importance established by these methods are due to Kühnau [1-13]. They have already been commented on in those sections. Also Gutljanskiĭ [1-3] as well as Gutljanskiĭ and Ščepetev [1, 2] obtained a number of results in this direction.

27. Basic applications in electrical engineering

Let us consider a charged particle (treated as a material point) of mass m and charge e. For the sake of simplicity we ignore relativistic effects. Denote by $\underline{v}$ the velocity of the particle, and by V and $\underline{A}$ the scalar and vector potential, respectively, of the field at the point where the particle actually appears. If we consider the field as given and make a variation of the particle trajectory, then the corresponding variational principle yields

$$(d/dt)\underline{p} = e\underline{E} + e\underline{v} \times \underline{B} \quad (\underline{\text{equations of motion}}),$$

where

(27.1) $\underline{p} = m\underline{v}$ (momentum)

(27.2) $\underline{E} = -(\partial/\partial t)\underline{A} - \operatorname{grad} V$ (intensity of the electric field),

(27.3) $\underline{B} = \operatorname{rot} \underline{A}$ (vector of the magnetic induction),

Eliminating $\underline{A}$ and V from (27.2) and (27.3) we get

(27.4) $\operatorname{rot} \underline{E} = -(\partial/\partial t)\underline{B}, \ \operatorname{div} \underline{B} = 0.$

The above result remains valid in the case of n charged particles of masses m_k and charges e_k, $k = 1,\dots,n$. With the corresponding meaning of $\underline{v}_k$, V_k, and $\underline{A}_k$, the variational principle yields

$$(d/dt)\underline{p} = \sum_{k=1}^{n} e_k(\underline{E}_k + \underline{v}_k \times \underline{B}_k) \quad (\underline{\text{equations of motion}})$$

where

(27.5) $\underline{p} = \sum_{k=1}^{n} m_k\underline{v}_k, \ \ \underline{E}_k = -\frac{\partial}{\partial t}\underline{A}_k - \operatorname{grad} V_k, \ \ \underline{B}_k = \operatorname{rot} \underline{A}_k.$

Then we consider the motion of particles as given and make a variation of the field itself, i.e. of the potentials. We assume that the charge is distributed continuously, and even more – that the mass density ζ as well as the charge density ρ are of the class C^2. It is clear that the discrete case also could be reduced to this one by means of the theory of distributions. A physical reasoning leads to the completion of the action integral applicable to a given field by the term

$$L = \tfrac{1}{2}\varepsilon_0\underline{\varepsilon}\,\underline{E}^2 - \tfrac{1}{2}\mu_0^{-1}\underline{\mu}^{-1}\underline{B}^2, \ \ \underline{E}^2 = [E_jE_k]_{j,k\le 3}, \ \ \underline{B}^2 = [B_jB_k]_{j,k\le 3},$$

where

$\underline{\varepsilon} = [\varepsilon_{j,k}]_{j,k\le 3}$ (tensor of electric permeability),

$\underline{\mu} = [\mu_{j,k}]_{j,k\le 3}$ (tensor of magnetic permeability),

$\varepsilon_{k,j} = \varepsilon_{j,k}, \ \mu_{k,j} = \mu_{j,k}$ for $k = 1,2,3$; $\det[\varepsilon_{j,k}] \ne 0$, $\det[\mu_{j,k}] \ne 0$, while ε_0 and μ_0 denote the electric and magnetic permeability of the vacuum, respectively. Now, the corresponding variational principle yields

(27.6) $\quad \mathrm{rot}\,\underline{H} = (\partial/\partial t)\underline{D} + \underline{j}, \ \mathrm{div}\,\underline{D} = \rho,$

where

(27.7) $\quad \underline{H} = \mu_0^{-1}\mu^{-1}\underline{B}$ (intensity of the magnetic field),

(27.8) $\quad \underline{D} = \varepsilon_0 \varepsilon \underline{E}$ (vector of the electric induction),

(27.9) $\quad \underline{j} = \rho \underline{v}$ (vector of the electric current-density).

Equations (27.4), (27.6), (27.8), and $\underline{B} = \mu_0 \mu \underline{H}$ are called the Maxwell equations.

Consider now the case of a plane, isotropic and inhomogeneous electrostatic field in a domain D, where there are no space charges, with (relative) electric permeability $\varepsilon = \varepsilon_{j,j}$ of the class C^1 (clearly. $\varepsilon_{j,k} = 0$ for $j \neq k$). By physical reasons we have always $\varepsilon > 1$. Then the function of forces U of the field is a solution of the system of equations

(27.10) $\quad U_x = (1/\varepsilon)V_y, \ U_y = -(1/\varepsilon)V_x,$

and the potential V is a solution of the equation

(27.11) $\quad \mathrm{div}(\varepsilon\,\mathrm{grad}\,V) = 0.$

This is a direct consequence of the Maxwell equations: (27.8) and the second equation in (27.6), and of the relation (27.2). To this case we can also reduce the situation where the plane field is anisotropic. Namely, we have to replace V and ε in (27.10) and (27.11) by V* and ε*, respectively, where V* and ε* are given by the following formulae due to R. Kühnau:

$$V_x^* = \varepsilon_*(V_x \cos u - V_y \sin u), \quad V_y^* = \frac{1}{\varepsilon_*}(V_x \sin u + V_y \cos u),$$

$$\varepsilon_* \cos^2 u + \frac{1}{\varepsilon_*}\sin^2 u = \frac{1}{\varepsilon^*}\varepsilon_{1,1}, \quad \varepsilon_* \sin^2 u + \frac{1}{\varepsilon_*}\cos^2 u = \frac{1}{\varepsilon^*}\varepsilon_{2,2},$$

$$\left(\varepsilon_* - \frac{1}{\varepsilon_*}\right)\cos u \sin u = \frac{1}{\varepsilon^*}\varepsilon_{1,2}, \quad \varepsilon^* = (\varepsilon_{1,1}\varepsilon_{2,2} - \varepsilon_{1,2}^2)^{\frac{1}{2}}$$

(for physical reasons $\varepsilon_{1,1}\varepsilon_{2,2} - \varepsilon_{1,2}^2 > 0$ and $\varepsilon_* > 0$).

From (27.10) and (27.11) we conclude that the function U fulfills an analogous equation to (27.11), namely

(27.12) $\quad \mathrm{div}[(1/\varepsilon)\,\mathrm{grad}\,U] = 0.$

A natural physical requirement for the lines of force $\{z: U(z) = \text{const}\}$ and equipotential lines $\{z: V(z) = \text{const}\}$ is equivalent to the requirement for the mapping $U + iV$ to be one-to-one. This mapping is called the complex potential of the field in question. It is clear that $U + iV$ is a qc mapping whose dilatation (the modulus of the complex dilatation) is equal to $(\varepsilon - 1)/(\varepsilon + 1)$. Now the physical meaning of irregular qc mappings in our context is also clear: they may serve for describing media with discontinuous electric permeability. Consequently, Theorems 2 and 3 are closely connected with the problem of existence and uniqueness of the complex potential of an electrostatic field, essentially solved by Parter [1].

In the case of a simply connected domain we usually consider a condenser being a finite closed Jordan domain D with two distinguished

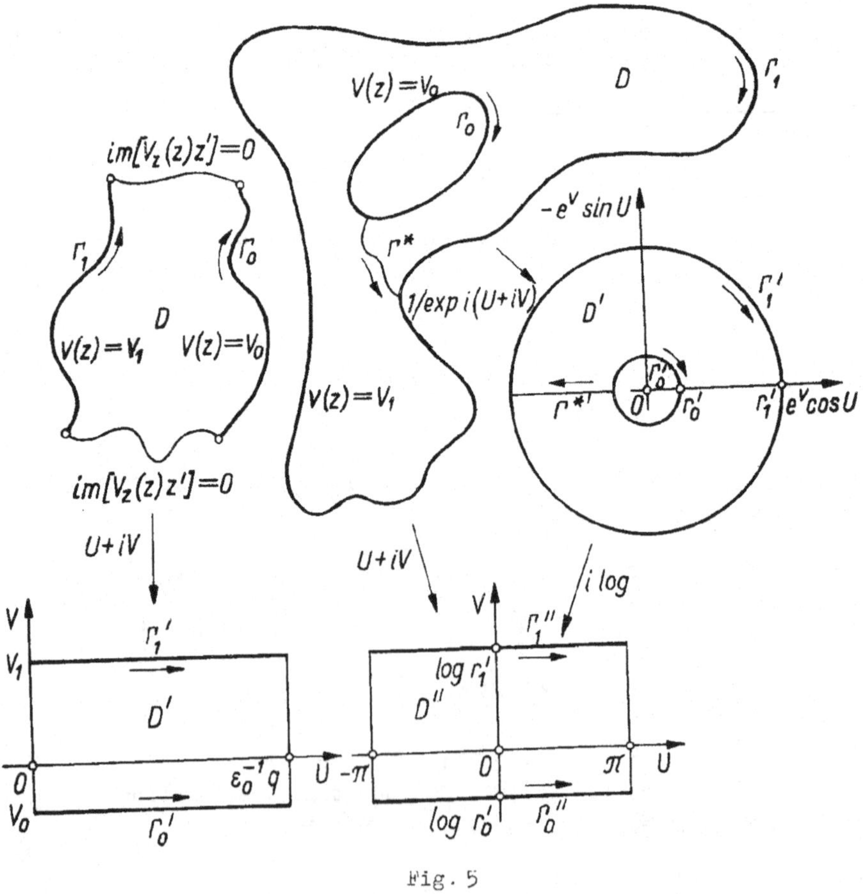

Fig. 5

disjoint boundary arcs Γ_0 and Γ_1 (the condenser plates), i.e. the quadrilateral $\mathbb{D} = (D, \Gamma_0, \Gamma_1)$. We suppose that $V(z) = V_0$ on Γ_0, $V(z) = V_1$ on Γ_1, V_0 and V_1 being constants, $V_0 \neq V_1$, and that the normal derivative $\operatorname{im}[V_z(z)z']$, i.e. $-V_y(z)x' + V_x(z)y'$ exists and vanishes at a.e. point z of $\operatorname{fr} D \setminus \Gamma_0 \setminus \Gamma_1$. Let the electric charges on Γ_0 and Γ_1 be $-q$ and q, respectively. Then $U + iV$ maps D onto a rectangular quadrilateral and the ratio $\varepsilon_0^{-1} q/(V_1 - V_0)$ of lengths of the corresponding sides, multiplied by ε_0, may be recognized as the capacity of D.

In the case of a doubly connected domain we usually consider a condenser being a finite closed domain D whose boundary consists of two disjoint Jordan curves Γ_0 and Γ_1 (the condenser plates). We suppose that $V(z) = V_0$ on Γ_0 and $V(z) = V_1$ on Γ_1, V_0 and V_1 being constants, $V_0 \neq V_1$. Now there is not any single-valued complex potential $U + iV$. From Theorem 3 or, more exactly, Corollary 4 we can only conclude the existence of a (single-valued) qc mapping $1/\exp i(U + iV)$ of D onto D', where D' is an annulus whose boundary circles have radii

$$r_0' = \exp[2\pi(\varepsilon_0/q)V_0], \quad r_1' = \exp[2\pi(\varepsilon_0/q)V_1].$$

Suppose now that $V_0 < V_1$, denote by $\Gamma^{*\prime}$ the segment joining the point $-r_0'$ to $-r_1'$, and let Γ^* be its preimage under the mapping $1/\exp i(U + iV)$. Then the function $U + iV$ exists as a single-valued function in $\operatorname{cl} D \setminus \Gamma^*$ and maps quasiconformally the condenser (in the preceding sense) $\mathbb{D} = (D \setminus \Gamma^*, \Gamma_0, \Gamma_1)$ onto $\mathbb{D}'' = (D'' \setminus (\operatorname{fr} D'' \setminus \Gamma_0'' \setminus \Gamma_1''), \Gamma_0'', \Gamma_1'')$, where D'' is a rectangle with sides of length $\log r_1' - \log r_0'$ and 2π. Of course we have to extend trivially the definition of qc mappings to the so-called semi-closed domains and to allow the simply connected finite closed domain appearing in the definition of a condenser in the preceding sense not to be Jordan. Now the ratio $2\pi/\log(r_1'/r_0')$ of lengths of the corresponding sides of D'', multiplied by ε_0, can be recognized as the capacity of D.

If, in particular, we consider a cylindrical condenser or a coaxial unistranded cable of radii $r_0 < r_1$ within a medium of a constant (relative) electric permeability ε, then the qc mapping $1/\exp(U + iV)$ admits the values $|z|^{1/\varepsilon} e^{i \arg z}$ for $r_0 \le |z| \le r_1$, so we have $r_0' = r_0^{1/\varepsilon}$, $r_1' = r_1^{1/\varepsilon}$ and, consequently, we arrive at the well known formula for the capacity:

$$C = 2\pi\varepsilon_0 / \log \frac{r_1'}{r_0'} = 2\pi\varepsilon_0\varepsilon / \log \frac{r_1}{r_0}.$$

Now we list some other physical models of the capacity defined as above, including for the sake of completeness also the previous model. Some of them were pointed out by R. Kühnau and J. Ławrynowicz. In all cases the capacity is considered apart from a multiplicative constant.

(a) V – potential of a constant electric field in a dielectric D between the plates Γ_0 and Γ_1, ε – electric permeability, $-\varepsilon\times\operatorname{grad} V$ = the vector of dielectric displacement, C – capacity of the condenser $\mathbb{D} = (D, \Gamma_0, \Gamma_1)$.

(b) V – as above, ε – electric susceptibility, $\varepsilon \operatorname{grad} V$ = the vector of dielectric polarization, C – entire electric susceptibility of the medium $\mathbb{D}$.

(c) V – potential of a constant electric field in a conductor insulated along $\operatorname{fr} D \setminus \Gamma_0 \setminus \Gamma_1$, ε – coefficient of electric conductivity, $-\varepsilon \operatorname{grad} V$ = the vector of electric current-density (the law of Ohm), C – intensity of electric current in the medium $\mathbb{D}$, divided by the voltage $V_1 - V_0$.

(d) V – potential of a constant magnetic field in a medium $\mathbb{D}$, ε – magnetic permeability, $-\varepsilon \operatorname{grad} V$ = the vector of magnetic induction, C – entire value of the vector of magnetic induction in $\mathbb{D}$, divided by the potential difference $V_1 - V_0$.

(e) V – as above, ε – magnetic susceptibility, $\varepsilon \operatorname{grad} V$ = the vector of magnetization, C – entire magnetic susceptibility of $\mathbb{D}$.

Interesting results within the model (a) for finitely connected domains have recently been obtained by Harrington [1] with the use of Schiffer's variational method. In particular, he has derived new inequalities of a dielectric-potential nature.

Already for a doubly connected D the model (a) shows how essential it is to consider multivalued functions: the complex potential $U + iV$ was not uniquely determined and this necessitated to use two evasions: at first we had to consider a uniquely determined qc mapping $f = 1/\exp i(U + iV)$, and then a uniquely determined restriction of $U + iV$ to $D \setminus \Gamma^*$. This problem becomes even more essential in the cases (d) and (e) of a magnetic field.

In domains with the vanishing resultant current density, e.g. in air, when neglecting dielectric displacement currents and moving charges, not only the complex magnetic potential, but also the scalar potential cannot be single-valued. One of the possible evasions is based on utilizing not a potential, but the <u>lines of force</u> (and, consequently, also the <u>functions of forces</u>); they can be defined as curves such that, at any point, their tangents have the same direction as the intensity $\underline{H}$ of the magnetic field at that point. In general, however, it is much more convenient to agree on considering a potential which is not single-valued. Let us look at the problem more closely, confining ourselves, for the sake of simplicity, to the plane case for media magnetically isotropic and homogeneous (we shall also be not concerned with the vector potential).

The magnetic potential V^* has to satisfy the equation $\operatorname{div}\operatorname{grad} V^* = 0$ which results from the Maxwell equations (27.6) and (27.8), and from the relation (27.9). Applying (27.6) again we conclude, by Green's formula, that the increment of V^* along the oriented boundary ∂D of an arbitrary domain D, $D \subset \mathbb{C}$, under the hypothesis that ∂D is rectifiable and composed of Jordan curves, has to be equal to the surface integral of the vector of the electric current-density (perpendicular to the plane in question) over the domain D, taken with the sign minus. This proves that we are unable to define V^* in D as a single-valued function.

In the case where the vector of the electric current-density vanishes apart from a finite number of measurable sets contained in

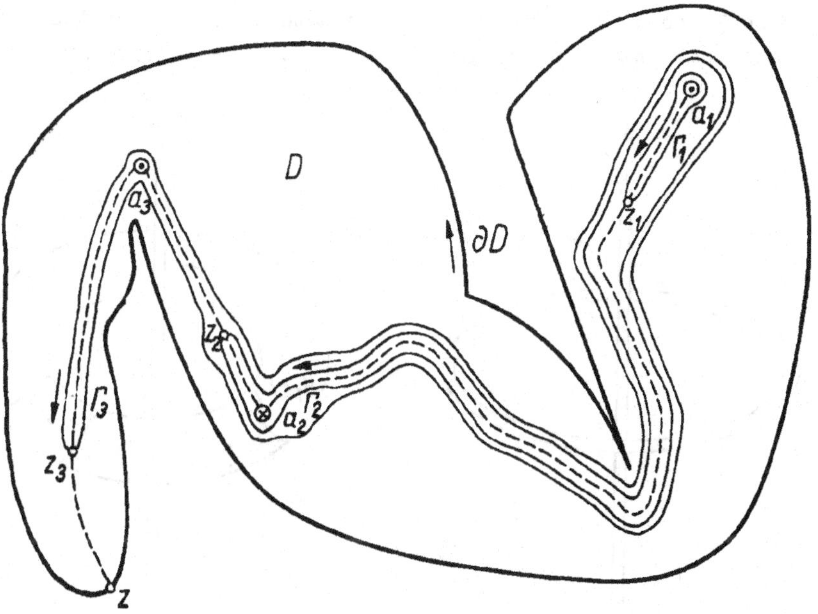

Fig. 6

discs of area small w.r.t. $|D|$, we can determine a potential as a single-valued function in the domain D which is suitably cut. As an example let us consider the situation visualized on Fig 6, where D is pierced perpendicularly to its plane at some points a_1, a_2, and a_3 by three conductors with current of intensity i_1, $-i_2$, and i_3, respectively. The increments of the potential V^* along the closed curves Γ_1, Γ_2, and Γ_3 (satisfying the regularity conditions analogous to ∂D), indicated on Fig. 6, should be $-i_1$, $i_2 - i_1$, and $-i_3 + i_2 - i_1$, respectively, and the last increment is equal to the increment along ∂D. Hence it is sufficient to introduce three arbitrary, fictitious rectifiable delimiting layers (or: magnetic shells) joining correspondingly e.g. the points a_1 to a_2, a_2 to a_3, and a_3 to z (an arbitrary point of ∂D), composed of nonintersecting Jordan curves, and then to classify the families of closed curves running within D according to the surrounded points of piercing a_1, a_2, a_3, and to the number of revolutions around a point in question. Thus we are naturally led to use the notions of the Riemann surface and of the fundamental group (cf. e.g. Springer [1], p. 85).

Finally let us add a few remarks on the lines of force of a magnetic field. On the basis of the Maxwell equations it can be proved that these lines cannot emanate or terminate at points of the field. This means that they either have to be closed, or have to run from in-

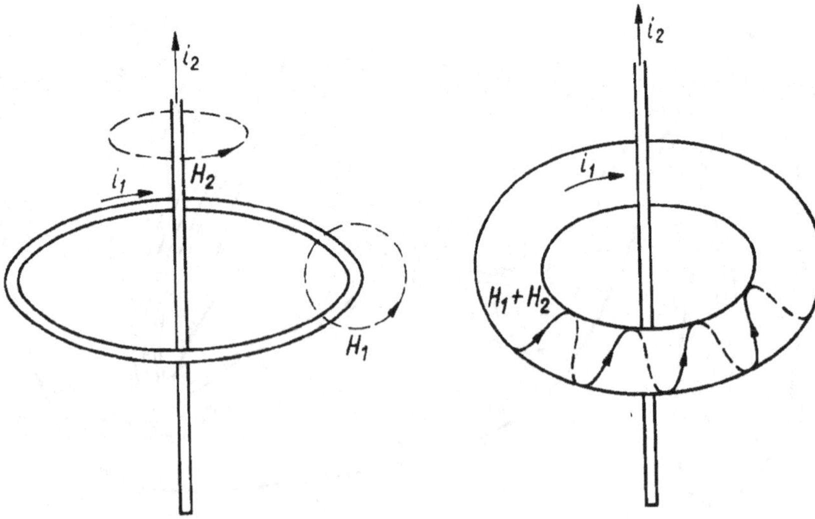

Fig. 7

finity to infinity, or have to cover some surface densely. The third case is often omitted what may lead to serious errors. A simple example of two electric currents producing a magnetic field with lines of force which do not emanate and terminate, and do not run from infinity to infinity is shown on Fig. 7 (cf. Tamm [1], pp. 239-242). Consequently, a magnetic field cannot be, in general, precisely visualized by means of the lines of force. Also the condition that the number of lines of force, intersecting a unit surface element perpendicular to them, is proportional to the intensity of the magnetic field of that element, what is often assumed, may fail to be true.

BIBLIOGRAPHY

AGARD, S. and F. W. GEHRING: [1] Angles and quasiconformal mappings, Proc. London Math. Soc. (3) 14A 1965 , 1-21.

AHLFORS, L. V.: [1] Conformal mapping, Lecture notes transcribed by R. Osserman at Oklahoma A. and M. College, 1951. [2] On quasiconformal mappings, J. Analyse Math. 3 (1954), 1-58 and 207-208. [3] Some remarks on Teichmüller's space of Riemann surfaces, Ann. Math. 74 (1961), 171-179. [4] Extension of quasiconformal mappings from two to three dimensions, Proc. Nat. Acad. Sci. U.S.A. 51 (1964), 768-771. [5] Lectures on quasiconformal mappings, Van Nostrand Co., Princeton 1966. [6] Complex analysis, 2. ed., McGraw-Hill Book Co., New York 1966. [7] Conformal invariants: Topics in geometric function theory, McGraw-Hill Book Co., New York 1973. [8] A remark on schlicht functions with quasiconformal extensions, Proceedings of the Symposium on Complex Analysis (London Math. Soc. Lecture Note Series 12), Cambridge Univ. Press, Cambridge 1974, pp. 7-10.

—— and L. BERS: [1] Riemann's mapping theorem for variable metrics, Ann. Math. 72 (1960), 385-404.

—— and A. BEURLING: [1] Invariants conformes et problèmes extrémaux, C. R. Dixième Congrès des Mathématiciens Scandinaves, Copenhague 1946, pp. 341-351. [2] Conformal invariants and function-theoretic null-sets, Acta Math. 83 (1950), 101-129.

—— and G. WEILL: [1] A uniqueness theorem for Beltrami equations, Proc. Amer. Math. Soc. 13 (1962), 975-978.

BECKER, J.: [1] Löwnersche Differentialgleichung und quasikonform fortsetzbare schlichte Funktionen, J. Reine Angew. Math. 255 (1972), 23-43. [2] Löwnersche Differentialgleichung und Schlichtheitskriterien, Math. Ann. 202 (1973), 321-335. [3] Über eine Golusinsche Ungleichung für quasikonform fortsetzbare schlichte Funktionen, Math. Z. 131 (1973), 177-182. [4] Über homöomorphe Fortsetzung schlichter Funktionen, Ann. Acad. Sci. Fenn. Ser. A I 578 (1973), 11 pp. [5] Über die Lösungsstruktur einer Differentialgleichung in der konformen Abbildung, J. Reine Angew. Math. 285 (1976), 66-74. [6] Conformal mappings with quasiconformal extensions, Aspects of contemporary complex analysis (Proc. NATO Adv. Study Inst., Univ. Durham, Durham 1979), Academic Press, London 1980, pp. 37-77.

—— and C. POMMERENKE: [1] Über die quasikonforme Fortsetzung schlichter Funktionen, Math. Z. 161 (1978), 69-80.

Bibliography

BEHNKE, H. and F. SOMMER: [1] Theorie der analytischen Funktionen einer komplexen Veränderlichen, 3. Aufl. (Grundlehren Math. Wissensch. 77), Springer-Verlag, Berlin - Heidelberg - New York 1965.

BELINSKIĬ, P. P.: [1] Solution of extremal problems in the theory of quasiconformal mappings by variational methods [in Russian], Sibirsk. Mat. Ž. 1 (1960), 303-330. [2] General properties of quasiconformal mappings [in Russian], Izdatelstvo "Nauka", Sibirskoe Otdelenie, Novosibirsk 1974.

—— and I. N. PESIN: [1] On the closure of the class of continously differentiable quasiconformal mappings [in Russian], Dokl. Akad. Nauk SSSR 102 (1955), 865-866.

BELTRAMI, E.: [1] Delle variabili complesse sopra una superficie qualunque, Ann. Mat. Pura Appl. (2) 1 (1867/8), 329-336.

BERS, L.: [1] On a theorem of Mori and the definition of quasiconformality, Trans. Amer. Math. Soc. 84 (1957), 78-84. [2] Quasiconformal mappings and Teichmüller's theorem, Analytic functions, Princeton Univ. Press, Princeton 1960, pp. 89-120. [3] The equivalence of two definitions of quasiconformal mappings, Comment. Math. Helv. 37 (1962), 148-154. [4] On moduli of Riemann surfaces, Lectures at Forschungsinstitut für Mathematik, Zürich 1964. [5] A new proof of a fundamental inequality for quasiconformal mappings, J. Analyse Math. 36 (1979), 15-30 (1980).

—— and L. NIRENBERG: [1] On a representation theorem for linear elliptic systems with discontinuous coefficients and its applications, Convegno internaz. equazioni lineari alle derivate parziali, Trieste 1954, Edizioni Cremonese, Roma 1955, pp. 111-140.

BOJARSKI, B.: [1] Homeomorphic solutions of Beltrami systems [in Russian], Dokl. Akad. Nauk SSSR 102 (1955), 661-664. [2] Generalized solutions of a system of differential equations of the first order and elliptic type with discontinuous coefficients [in Russian], Mat. Sb. N. S. 43 (1957), 451-503.

—— and T. IWANIEC: [1] Quasiconformal mappings and nonlinear elliptic equations in two variables I-II, Bull. Acad. Polon. Sci. Sér. Sci. Math. Astronom. Phys. 22 (1974), 473-484.

CALDERÓN, A. P. and A. ZYGMUND: [1] On the theorem of Hausdorff-Young and its extensions, Contributions to Fourier Analysis (Annals of Mathematics Studies 25), Princeton Univ. Press, Princeton 1950, pp. 145-165. [2] On the existence of certain singular integrals, Acta Math. 88 (1952), 88-139. [3] On singular integrals, Amer. J. Math. 78 (1956), 289-309.

CARATHÉODORY, C.: [1] Über die Begrenzung einfachzusammenhängenden Gebiete, Math. Ann. 73 (1913), 323-370. [2] Vorlesungen über reelle Funktionen, Verlag von B. G. Teubner, Leipzig - Berlin 1918.

CHENG Bao-long: [1] Approximate representations and extremal problems of ε-quasiconformal mappings [in Chinese], Acta Math. Sinica 14 (1964), 212-217; English translation appeared in Chinese Math. 5 (1964), 233-248.

CODDINGTON, E. A. and N. LEVINSON: [1] Theory of ordinary differential equations, McGraw-Hill Book Co., New York - Toronto - London 1955.

COLLINGWOOD, E. F. and A. J. LOHWATER: [1] The theory of cluster sets (Cambridge Tracts in Math. and Math. Phys. 56), Cambridge Univ. Press, Cambridge 1956.

Bibliography

COURANT, R.: [1] Über direkte Methode in der Variationsrechnung und über verwandte Fragen, Math. Ann. 97 (1927), 711-736. [2] Dirichlet's principle, conformal mapping, and minimal surfaces, Interscience Publishers, New York - London 1950.

DITTMAR, B.: [1] Ein neuer Existenzbeweis für quasikonforme Abbildungen mit vorgegebener komplexer Dilatation, Analytic Functions, Kozubnik 1979, Proceedings (Lecture Notes in Mathematics 798), Springer-Verlag, Berlin - Heidelberg - New York 1980, pp. 148-154.

DUREN, P. L. and O. LEHTO: [1] Schwarzian derivatives and homeomorphic extensions, Ann. Acad. Sci. Fenn. Ser. A I 477 (1970), 11 pp.

FAIT, M., J. KRZYŻ, and J. ZYGMUNT: [1] Explicit quasiconformal extensions for some classes of univalent functions, Comment. Math. Helv. 51 (1976), 279-285.

GAUSS, C. F.: [1] Allgemeine Auflösung der Aufgabe, die Theile einer gegebenen Fläche auf einer anderer gegebenen Fläche so abzubilden, daß die Abbildung dem Abgebildeten in der kleinsten Theilen ähnlich wird, Carl Friedrich Gauß Werke, Bd. IV, Dieterische Universitäts-Druckerei, Göttingen 1880.

GEHRING, F. W.: [1] The definitions and exceptional sets for quasiconformal mappings, Ann. Acad. Sci. Fenn. Ser. A I 281 (1960), 28 pp. [2] Definitions for a class of plane quasiconformal mappings, Nagoya Math. J. 29 (1967), 175-184. [3] Quasiconformal mappings which hold the real axis pointwise fixed, Mathematical Essays dedicated to A. J. MacIntyre, Ohio Univ. Press, Athens 1970, pp. 145-148. [4] Inequalities for condensers, hyperbolic capacity, and extremal lengths, Michigan Math. J. 18 (1971), 1-20. [5] The L^p-integrability of the partial derivatives of a quasiconformal mapping, (a) Bull. Amer. Math. Soc. 79 (1973), 465-466, (b) Acta Math. 130 (1973), 265-277.

—— and O. LEHTO: [1] On the total differentiability of functions of a complex variable, Ann. Acad. Sci. Fenn. Ser. A I 272 (1959), 9 pp.

—— and E. REICH: [1] Area distortion under quasiconformal mappings, Ann. Acad. Sci. Fenn. Ser. A I 388 (1966), 15 pp.

—— and J. VÄISÄLÄ: [1] On the geometric definition for quasiconformal mappings, Comment. Math. Helv. 36 (1961), 19-32.

GÖKTÜRK, Z.: [1] Estimates for univalent functions with quasiconformal extensions, Ann. Acad. Sci. Fenn. Ser. A I 589 (1974), 21 pp.

GOL'DŠTEĬN, V. M.: [1] Degree of summability of generalized derivatives of plane quasiconformal homeomorphisms [in Russian], Dokl. Akad. Nauk SSSR 250 (1980), 18-21.

GOLUSIN, G. M.: [1] On the parametric representation of functions univalent in a ring [in Russian], Math. Sb. N. S. 29 (1951), 469-476. [2] Geometrische Funktionentheorie, VEB Deutscher Verlag der Wissenschaft, Berlin 1957.

GOURSAT, É.: [1] Cours d'analyse mathématique, 3. éd., Gauthier-Villars, Paris 1917.

GRAVES, L. M.: [1] The theory of functions of real variables, McGraw-Hill Book Co., New York - London 1946.

GRÖTZSCH, H.: [1] Über die Verzerrung bei schlichten nichtkonformen Abbildungen und über damit zusammenhängende Erweiterung des Picardschen Satzes, Ber. Verh. Sächs. Akad. Wiss. Leipzig Math.-Phys. Kl. 80 (1928), 503-507. [2] Über möglichst konforme Abbildungen von schlichten Bereichen, ibid. 84 (1932), 114-120.

GRUNSKY, H.: [1] Neue Abschätzungen zur konformen Abbildung ein und mehrfach zusammenhängender Bereiche, Schr. Math. Sem. Inst. Angew. Math. Univ. Berlin 1 (1932), 93-140.

GUTLJANSKIĬ, V. Ja.: [1] The area principle for a certain class of quasiconformal mappings (in Russian), Dokl. Akad. Nauk SSSR 212 (1973), 540-543. [2] The method of variations for univalent analytic functions with a quasiconformal extension [in Russian], (a) ibid. 236 (1977), 1045-1048, (b) Sibirsk. Mat. Ž. 21 (1980), 61-78, 237. [3] Distortion theorems for univalent analytic functions with a quasiconformal extension [in Russian], Dokl. Akad. Nauk SSSR 240 (1978), 515-517.

—— and V. A. Ščepetev: [1] A generalized area theorem for a certain class of q-quasiconformal mappings [in Russian], Dokl. Akad. Nauk SSSR 218 (1974), 509-512. [2] Exact estimates of the modulus of a univalent analytic function with a quasiconformal extension [in Russian], Akad. Nauk Ukrain. SSR Inst. Mat. Preprint 1979, no. 13, 28 pp.

HADAMARD, J.: [1] Mémoire sur le problème d'analyse relatif à l'équilibre des plaques élastiques encastrées, Mém. Acad. Sci. Inst. France (2) 33,4 (1907), 128 pp.

HAMILTON, R. S.: [1] Extremal quasiconformal mappings with prescribed boundary values, Trans. Amer. Math. Soc. 138 (1969), 339-406.

HARRINGTON, A. N.: [1] Some extremal problems in conformal and quasiconformal mapping, Michigan Math. J. 27 (1980), 95-116.

HERSCH, J.: [1] Contribution à la théorie des fonctions pseudo-analytiques, Comment. Math. Helv. 30 (1956), 1-19.

—— and A. PFLUGER: [1] Généralisation du lemme de Schwarz et du principe de la mesure harmonique pour les fonctions pseudo-analytiques, C. R. Acad. Sci. Paris 234 (1952), 43-45.

HÜBNER, O.: [1] Remarks on a paper by Ławrynowicz on quasiconformal mappings, Bull. Acad. Polon. Sci. Sér. Sci. Math. Astronom. Phys. 18 (1970), 183-186.

JULIA, G.: [1] Sur une équation aux dérivées fonctionnelles liée à la représentation conforme, Ann. Sci. École Norm. Sup. (3) 32 (1922), 1-28.

KELINGOS, J. A.: [1] Characterizations of quasiconformal mappings in terms of harmonic and hyperbolic measure, Ann. Acad. Sci. Fenn. Ser. A I 368 (1965), 16 pp. [2] Distortion of hyperbolic area under quasiconformal mappings, Duke Math. J. 41 (1974), 127-139.

KOMATU, Y.: [1] Untersuchungen über konforme Abbildung zweifachzusammenhängender Bereiche, Proc. Phys.-Math. Soc. Japan 25 (1943), 1-42.

KRUŠKAL, S. L.: [1] The method of variations in the theory of quasiconformal mappings of closed Riemann surfaces [in Russian], Dokl. Akad. Nauk SSSR 157 (1964), 781-783. [2] The variation of a quasiconformal mapping of an annulus [in Russian], Sibirsk. Mat. Ž. 5 (1964), 236-239. [3] On a Teichmüller theorem on extremal quasiconformal mappings [in Russian], ibid. 8 (1967), 313-332. [4] On the theory of extremal problems for quasiconformal mappings of closed Riemann surfaces [in Russian], Dokl. Akad. Nauk SSSR 171 (1966), 784-787. [5] Mappings ε-quasiconformal in the mean [in Russian], Sibirsk. Mat. Ž. 8 (1967), 798-806. [6] Some extremal problems for univalent analytic functions [in Russian], Dokl. Akad. Nauk SSSR

182 (1968), 754-757. [7] Variational methods in the theory of quasiconformal mappings [in Russian], Novosibirskiĭ Gosudarstvennyĭ Universitet, Novosibirsk 1974. [8] Quasiconformal mappings and Riemann surfaces [in Russian], Izdatelstvo "Nauka", Sibirskoe Otdelenie, Novosibirsk 1975.

KRZYŻ, J.: [1] On an extremal problem of F. W. Gehring, Bull. Acad. Polon. Sci. Sér. Sci. Math. Astronom. Phys. 16 (1968), 99-101. [2] Problems in complex variable theory, American Elsevier Publ. Co. and PWN - Polish Scientifical Publishers, Warszawa 1971. [3] Convolution and quasiconformal extension, Comment. Math. Helv. 51 (1976), 99-104. [4] Über schlichte quasikonform fortsetzbare Funktionen, Komplexe Analysis und ihre Anwendung (Tagung), Wiss. Beitr. Martin-Luther-Univ. Halle-Wittenberg M 27 (1977), 33-35.

—— and J. ŁAWRYNOWICZ: [1] Quasiconformal mappings of the unit disc with two invariant points, Michigan Math. J. 14 (1967), 487-492 and 15 (1968), 506.

—— and A. K. SONI: [1] Close-to-convex functions with quasiconformal extension, in: Analytic Functions, Błażejewko 1982, Proceedings (Lecture Notes in Mathematics), Springer-Verlag, Berlin - Heidelberg - New York, to appear.

KÜHNAU, R.: [1] Einige Extremalprobleme bei differentialgeometrischen und quasikonformen Abbildungen I-II, Math. Z. 94 (1966), 178-192 and 107 (1968), 307-318. [2] Wertannahmeprobleme bei quasikonformen Abbildungen mit ortsabhängiger Dilatationsbeschränkung, Math. Nachr. 40 (1969), 1-11. [3] Koeffizientenbedingungen bei quasikonformen Abbildungen, Ann. Univ. Mariae Curie-Skłodowska Sect. A 22/23/24 (1968/9/70), 105-111. [4] Bemerkungen zu den Grunskyschen Gebieten, Math. Nachr. 44 (1970), 285-293. [5] Verzerrungssätze und Koeffizientenbedingungen von Grunskyschen Typ für quasikonforme Abbildungen, ibid. 48 (1971), 77-105. [6] Zum Koeffizientenproblem bei den quasikonform fortsetzbaren schlichten konformen Abbildungen, ibid. 55 (1973), 225-231. [7] Eine Klasse nichtschlichter konformer Abbildungen mit einer schlichten quasikonformen Fortsetzung, ibid. 59 (1974), 261-263. [8] Zur quasikonformen Fortsetzbarkeit schlichter konformer Abbildungen, Bull. Soc. Sci. Lettres Łódź 24,6 (1974), 4 pp. [9] Extremalprobleme bei quasikonformen Abbildungen mit kreisringweisekonstanter Dilatationsbeschränkung, Math. Nachr. 66 (1975), 269-282. [10] Eine Verschärfung des Koebeschen Viertelsatzes für quasikonform fortsetzbare Abbildungen, Ann. Acad. Sci. Fenn. Ser. A I Math. 1 (1975), 77-83. [11] Zur Methode der Randintegration bei quasikonformen Abbildungen, Ann. Polon. Math. 31 (1975/6), 269-289. [12] Verzerrungsaussagen bei quasikonformen Abbildungen mit ortsabhängiger Dilatationsbeschränkung und ein Extremalprinzip der Elektrostatik in inhomogenen Medien, Comment. Math. Helv. 53 (1978), 408-428. [13] Schlichte konforme Abbildungen auf nichtüberlappende Gebiete mit gemeinsamer quasikonformer Fortsetzung, Math. Nachr. 86 (1978), 175-180. [14] Zur Modulânderung eines Vierecks bei quasikonformer Abbildung, Math. Nachr. 93 (1979), 249-258. [15] Über die Werte des Doppelverhältnisses bei quasikonformer Abbildung, ibid. 95 (1980), 237-251.

—— and H. BAUMGARTEN: [1] Die Koeffizientenbedingungen von Grunskyschen Typ für quasikonforme Abbildungen mit längs zweier Kreise springender Dilatationsschränke, Math. Nachr. 92 (1979), 117-127.

—— and H. BLAAR: [1] Kriterien für quasikonforme Fortsetzbarkeit konformer Abbildungen eines Kreisringes bzw. Inneren oder Äußeren einer Ellipse, Math. Nachr. 91 (1979), 183-196.

Bibliography

—— and B. DITTMAR: [1] Einige Folgerungen aus den Koeffizientenbedingungen vom Grunskyschen Typ für schlichte quasikonform fortsetzbare Abbildungen, Math. Nachr. 66 (1975), 5-16.

—— and E. HOY: [1] Abschätzung des Wertebereichs einiger Funktionale bei quasikonformen Abbildungen, Bull. Soc. Sci. Lettres Łódź 29,4 (1979), 9 pp.

—— and W. NISKE: [1] Abschätzung des dritten Koeffizienten bei den quasikonform fortsetzbaren schlichten Funktionen der Klasse S, Math. Nachr. 78 (1977), 185-192.

—— and B. THÜRING: [1] Berechnung einer quasikonformen Extremalfunktion, Math. Nachr. 79 (1977), 99-113.

—— and J. TIMMEL: [1] Asymptotische Koeffizientenabschätzungen für die quasikonform fortsetzbaren Abbildungen der Klasse S bzw. Σ, Math. Nachr. 91 (1979), 357-362.

LAVRENTIEFF, M. A.: [1] Sur une méthode géométrique dans la représentation conforme, Atti del Congresso internazionale dei matematici, Bologna, 3-10 settembre 1928, Comunicazioni sezione I, 1928 (C-D), Zanichelli, Bologna 1930, pp. 241-242. [2] Sur une classe de représentations continues, Mat. Sb. 48 (1935), 407-424.

ŁAWRYNOWICZ, J.: [1] On the parametrization of quasiconformal mappings in an annulus, Ann. Univ. Mariae Curie-Skłodowska Sect. A 18 (1964), 23-52. [2] Some parametrization theorems for quasiconformal mappings in an annulus, Bull. Acad. Polon. Sci. Sér. Sci. Math. Astronom. Phys. 15 (1967), 319-323. [3] Quasiconformal mappings of the unit disc near to the identity, ibid. 16 (1968), 771-777. [4] On the class of quasi-conformal mappings with invariant boundary points I-II, Ann. Polon. Math. 21 (1969), 309-347. [5] On arbitrary homotopies in parametrization theorems for quasiconformal mappings, Bull. Acad. Polon. Sci. Ser. Sci. Math. Astronom. Phys. 20 (1972), 733-737. [6] On the parametrization of quasiconformal mappings with invariant boundary points in the unit disc, ibid. 20 (1972), 739-744. [7] On the parametrization of quasiconformal mappings with invariant boundary points in an annulus, Comment. Math. Helv. 47 (1972), 213-219. [8] Parametrization and boundary correspondence for Teichmüller mappings in an annulus, Romanian-Finnish Seminar on Complex Analysis, Proceedings, Bucharest 1976 (Lecture Notes in Mathematics 743), Springer-Verlag, Berlin - Heidelberg - New York 1979, pp. 165-183. [9] Variationsrechnung und Anwendungen, Springer-Verlag, Berlin - Heidelberg - New York, to appear.

LEHTO, O.: [1] Remarks on the integrability of the derivatives of quasiconformal mappings, Ann. Acad. Sci. Fenn. Ser. A I 371 (1965), 8 pp. [2] Entwicklung der Theorie quasikonformer Abbildungen, Mitt. Math. Gesellsch. DDR 1970, No. 3-4, 36-47. [3] Schlicht functions with a quasiconformal extension, Ann. Acad. Sci. Fenn. Ser. A I 500 (1971), 10 pp. [4] Conformal mappings and Teichmüller spaces, Notes of lectures given at the Technion, April-May 1973, Technion - IIT, Haifa 1973, 58 pp. [5] Quasiconformal mappings in the plane, Lectures on quasiconformal mappings (Dept. Math., Univ. Maryland, Lecture Note, No. 14), Dept. Math. Univ. Maryland, College Park, Md., 1975, 43 pp. [6] Quasiconformal mappings and singular integrals, Symposia Mathematica XVIII, Istituto Nazionale di Alta Matematica, Roma, Academic Press, London - New York 1976, pp. 429-453. [7] Analytic families of quasiconformal mappings, Ann. Polon. Math. 33 (1976), 57-62. [8] On univalent functions with quasiconformal extensions over the boundary, J. Analyse Math. 30 (1976), 349-354.

—— and O. TAMMI: [1] Area method and univalent functions with quasiconformal extensions, Ann. Acad. Sci. Fenn. Ser. A I Math. 2 (1976), 307-313.

—— and K. I. VIRTANEN: [1] Quasikonforme Abbildungen (Grundlehren Math. Wissensch. 126), Springer-Verlag, Berlin - Heidelberg - New York 1965. [2] Quasiconformal mappings in the plane, 2. ed. (Grundlehren Math. Wissensch. 126), Springer-Verlag, Berlin - Heidelberg - New York 1973.

——, K. I. VIRTANEN, and J. VÄISÄLÄ: [1] Contributions to the distortion theory of quasiconformal mappings, Ann. Acad. Sci. Fenn. Ser. A I 273 (1959), 14 pp.

LICHTENSTEIN, L.: [1] Zur Theorie der konformen Abbildung. Konforme Abbildung nichtanalytischer singularitätenfreier Flächenstücke auf eine Gebiete, Bull. Acad. Sci. Cracovie 1916, 192-217.

LÖWNER, K.: [1] Untersuchungen über schlichte konforme Abbildungen des Einheitskreises I, Math. Ann. 89 (1923), 103-121.

MORI, A.: [1] On quasi-conformality and pseudo-analyticity, Trans. Amer. Math. Soc. 84 (1957), 56-77.

MORREY, C. B., Jr.: [1] On the solutions of quasi-linear elliptic partial differential equations, Trans. Amer. Math. Soc. 43 (1938), 126-166. [2] Multiple integrals in the calculus of variations (Grundlehren Math. Wissensch. 130), Springer-Verlag, Berlin - Heidelberg - New York 1966.

NEWMAN, M. H. A.: [1] Elements of the topology of plane sets of points, Cambridge Univ. Press, Cambridge 1951.

PARTER, S. V.: [1] On mappings of multiply connected domains by solutions of partial differential equations, Comm. Pure Appl. Math. 13 (1960), 167-182.

PESIN, I. N.: [1] Metric properties of Q-quasiconformal mappings [in Russian], Mat. Sb. N. S. 40 (1956), 281-294.

PETROVSKI, I. G.: [1] Ordinary differential equations, Prentice-Hall, Inc., Englewood Cliffs, N. J., 1966.

PFLUGER, A.: [1] Quasikonforme Abbildungen und logaritmische Kapazität, Ann. Inst. Fourier Grenoble 2 (1951), 69-80. [2] Über die Äquivalenz der geometrischen und analytischen Definition quasikonformer Abbildungen, Comment. Math. Helv. 33 (1959), 23-33.

PÓLYA, G. and G. SZEGÖ: [1] Aufgaben und Lehrsätze aus der Analysis I-II (Grundlehren Math. Wissensch. 19-20), Springer-Verlag, Berlin - Göttingen - Heidelberg 1925.

POMMERENKE, C.: [1] Univalent functions (Mathematische Lehrbücher 25), Vanderhoeck und Ruprecht, Göttingen 1975.

REICH, E.: [1] On a characterization of quasiconformal mappings, Comment. Math. Helv. 37 (1962), 44-48. [2] Some estimates for the two-dimensional Hilbert transform, J. Analyse Math. 18 (1967), 279-293. [3] Quasiconformal mappings of the disk with given boundary values, Advances in complex function theory (Proc. Sem., Univ. Maryland, College Park, Md., 1973), (Lecture Notes in Mathematics 505), Springer-Verlag, Berlin - Heidelberg - New York 1976, pp. 101-137. [4] On the decomposition of a class of plane quasiconformal mappings, Comment. Math. Helv. 53 (1978), 15-27. [5] Quasiconformal mappings with prescribed boundary values and a dilatation bound, Arch. Rational Mech. Anal. 68 (1978), 99-112. [6] On the uniqueness problem for extremal quasiconformal mappings with prescribed boundary values, in: Complex analysis, Joensuu 1978, Proceedings (Lecture Notes in Mathematics 747), Springer-Verlag, Berlin - Heidelberg - New York 1979, pp. 314-320. [7] Uniqueness of Hahn-Banach extensions from certain spaces of analytic functions, Math. Z. 167 (1979), 81-89.

—— and K. STREBEL: [1] On quasiconformal mappings which keep the boundary points fixed, Trans. Amer. Math. Soc. 138 (1939), 211-222. [2] Einige Klassen Teichmüllerscher Abbildungen, die die Randpunkte festhalten, Ann. Acad. Sci. Fenn. Ser. A I 457 (1970), 19 pp. [3] Extremal quasiconformal mappings with given boundary values, (a) Bull. Amer. Math. Soc. 79 (1973), 488-490, (b) Contributions to analysis (a collection of papers dedicated to Lipman Bers), Academic Press, New York 1974, pp. 375-391.

RENELT, H.: [1] Behandlung von Abbildungsproblemen der quasikonformen Abbildung mittels direkter Variationsmethoden, Inaugurationsdissertation zur Erlangung der Doktorwürde der Fakultät für Naturwissenschaften des Wissenschaftliches Rates der Martin-Luther-Universität Halle-Wittenberg, Halle an der Saale, Mai 1971, XII + 61 pp. [2] Über quasikonforme Abbildungen mehrfachzusammenhängender Gebiete durch Lösungen elliptischer Differentialgleichungssysteme, Ann. Univ. Mariae Curie-Skłodowska Sect. A 22/23/24 (1968/9/70), 155-160. [3] Modifizierung und Erweiterung einer Schifferschen Variationsmethode für quasikonforme Abbildungen, Math. Nachr. 55 (1973), 353-379. [4] Über Extremalprobleme für schlichte Lösungen elliptischer Differentialgleichungssysteme, Comment. Math. Helv. 54 (1979), 17-41.

RIESZ, M.: [1] Sur les maxima des formes bilinéaires et sur les fonctionnelles lineaires, Acta Math. 49 (1926), 465-497. [2] Sur les fonctions conjugées, Math. Z. 27 (1927), 218-244.

SAKS, S.: [1] Théorie de l'intégrale (Monografje Matematyczne 2), z subwencji Funduszu Kultury narodowej, Warszawa - Lwów 1933. [2] Theory of the integral (Monografie Matematyczne 7), z subwencji Funduszu Kultury Narodowej, Warszawa - Lwów 1937.

—— and A. ZYGMUND: [1] Analytic functions, 2. ed. (Monografie Matematyczne 28), PWN - Polish Scientifical Publishers, Warszawa 1971.

SCHAEFFER, A. C. and D. C. SPENCER: [1] Coefficient regions for schlicht functions (Amer. Math. Soc. Colloquium Publications 25), Amer. Math. Soc., New York 1950.

SCHIFFER, M.: [1] A method of variation within the family of simple functions, Proc. London Math. Soc. 44 (1938), 432-449. [2] A variational method for univalent quasiconformal mappings, Duke Math. J. 33 (1966), 395-411.

—— and G. SCHOBER: [1] An extremal problem for the Fredholm eigenvalues, Arch. Rational Mech. Anal. 44 (1971), 83-92 and 46 (1972), 394. [2] Coefficient problems and generalized Grunsky inequalities for schlicht functions with quasiconformal extensions, ibid. 60 (1976), 205-228. [3] Representation of fundamental solutions for generalized Cauchy-Riemann equations by quasiconformal mappings, Ann. Acad. Sci. Fenn. Ser. A I Math. 2 (1976), 501-531. [4] A variational method for general families of quasiconformal mappings, J. Analyse Math. 34 (1978), 240-264 (1979). [5] An application of the calculus of variations for general families of quasiconformal mappings, in: Complex analysis, Joensuu 1978, Proceedings (Lecture Notes in Mathematics 747), Springer-Verlag, Berlin - Heidelberg -

New York 1979, pp. 349-357. [6] The dielectric Green's function and quasiconformal extensions, J. Analyse Math. 36 (1979), 233-243 (1980).

SCHOBER, G.: [1] Univalent functions - Selected topics (Lecture Notes in Mathematics 478), Springer-Verlag, Berlin - Heidelberg - New York 1975. [2] Coefficients of inverses of univalent functions with quasiconformal extensions, Kōdai Math. J. 2 (1979), 411-419.

ŠERETOV, V. G.: [1] On the theory of extreme quasiconformal mappings [in Russian], Mat. Sb. N. S. 107 (1978), 146-158. [2] Extremal quasiconformal mappings with a given boundary correspondence [in Russian], Sibirsk. Mat. Ž. 19 (1978), 942-952, 957. [3] Locally extremal quasiconformal mappings [in Russian], Dokl. Akad. Nauk SSSR 250 (1980), 1338-1340.

SHAH Tao-shing: [1] Parametrical representation of quasiconformal mappings [in Russian], Science Record N. S. 3 (1959), 400-407.

—— and FAN Le-le: [1] On the modulus of quasi-conformal mappings, Science Record N. S. 4 (1960), 323-328. [2] On the parametric representation of quasiconformal mappings, Scientia Sinica 2 (1962), 149-162.

SHIBATA, K.: [1] On the defining properties of Teichmüller map, Osaka J. Math. 14 (1977), 95-109.

SPRINGER, G.: [1] Introduction to Riemann surfaces, Addison-Wesley Publ. Co., Reading, Mass., 1957.

STEIN, P.: [1] On a theorem of M. Riesz, J. London Math. Soc. 8 (1933), 242-247.

STREBEL, K.: [1] On the maximal dilatation of quasiconformal mappings, Proc. Amer. Math. Soc. 6 (1955), 903-909. [2] Zur Frage der Eindeutigkeit extremaler quasikonformer Abbildungen des Einheitskreises I-II, Comment. Math. Helv. 36 (1962), 306-323 and 39 (1964), 77-89. [3] Über quadratische Differentiale mit geschlossenen Trajektorien und extremale quasikonforme Abbildungen, Festband zum 70. Geburtstag von Prof. Rolf Nevanlinna, Springer-Verlag, Berlin - Heidelberg - New York 1966, pp. 105-127. [4] Bemerkungen über quadratische Differentiale mit geschlossenen Trajektorien, Ann. Acad. Sci. Fenn. Ser. A I 405 (1967), 12 pp. [5] On quadratic differentials and extremal quasiconformal mappings, Lecture Notes, Univ. of Minnesota, Minneapolis 1967. [6] Ein Konvergenzsatz für Folgen quasikonformer Abbildungen, Comment. Math. Helv. 44 (1969), 469-475. [7] On the existence of extremal quasiconformal mappings, J. Analyse Math. 30 (1976), 464-480. [8] On quasiconformal mappings of open Riemann surfaces, Comment. Math. Helv. 53 (1978), 301-321.

TAARI, O.: [1] Charakterisierung der Quasikonformität mit Hilfe der Winkelverzerrung, Ann. Acad. Sci. Fenn. Ser. A I 390 (1966), 43 pp.

TAMARKIN, J. D. and A. ZYGMUND: [1] Proof of a theorem of Thorin, Bull. Amer. Math. Soc. 50 1944 , 279-282.

TAMM, I. E.: [1] Foundations of the theory of electricity [in Russian], 8. ed., Izdatelstvo "Nauka", Moscow 1966.

TANAKA, H.: [1] On a theorem of Koebe, in: Quasiconformal mappings and Riemann surfaces, Proceedings of a Symposium held at the Research Institute for Mathematical Sciences, Kyoto Univ., Kyoto 1979, pp. 16-24.

TEICHMÜLLER, O.: [1] Extremale quasikonforme Abbildungen und quadratische Differentiale, Abh. Preuss. Akad. Wiss. Math.-Naturwiss. Kl.

22 (1940), 197 pp. [2] Ein Verschiebungssatz der quasikonformen Abbildungen, Deutsche Mathematik 7 (1944), 336-343. [3] Bestimmung der extremalen quasikonformen Abbildungen bei geschlossenen orientierten Riemannschen Flächen, Abh. Preusss Akad. Wiss. Math.-Naturwiss. Kl. 4 (1943), 42 pp. (1944).

THORIN, G. O.: [1] An extension of a convexity theorem due to M. Riesz, Kungl. Fysiografiska Sällskapets i Lund Förhandingar 8 (1939), no. 14. [2] Convexity theorems generalizing those of M. Riesz and Hadamard with some applications, Medd. Lunds Univ. Mat. Sem. 9 (1948), 1-58.

TRICOMI, F.: [1] Equazioni integrali contenenti il valor principale di un integrale doppio, Math. Z. 27 (1928), 87-133.

VÄISÄLÄ, J.: [1] On quasiconformal mappings in space, Ann. Acad. Sci. Fenn. Ser. A I 298 (1961), 36 pp.

VEKUA, I. N.: [1] The problem of reduction to canonical form of diferential forms of elliptic type and the generalized Cauchy-Riemann system [in Russian], Dokl. Akad. Nauk SSSR 100 (1955), 197-200. [2] Generalized analytic functions (Internat. Ser. Pure Appl. Math. 25), Pergamon Press, Oxford - London - New York - Paris 1962.

WANG Chuan-fang: [1] On the precision of Mori's theorem in Q-mapping, Science Record N. S. 4 (1960), 329-333.

ZAJĄC, J.: [1] On the parametrization of Teichmüller mappings in an annulus, Bull. Soc. Sci. Lettres Łódź 27,1 (1977), 8 pp. [2] The Ahlfors class N and its connection with Teichmüller quasiconformal mappings of an annulus, Ann. Univ. Mariae Curie-Skłodowska Sect. A 32 (1978), 155-162.

ZYGMUND, A.: [1] Trigonometrical series (Monografje Matematyczne 5), z subwencji Funduszu Kultury Narodowej, Warszawa - Lwów 1935. [2] Trigonometric series I-II, Cambridge Univ. Press, Cambridge 1959.

LIST OF SYMBOLS AND ABBREVIATIONS

AUTHOR INDEX [Numbers in square brackets refer to papers]

Author index

SUBJECT INDEX

Subject index

Subject index

ERRATA to LECTURE NOTES IN MATHEMATICS 978

80^7 [page 80, line 7 from above] $\varphi(\ ,t)$ repl. by $\mu(\ ,t)$
[$\varphi(\ ,t)$ should be replaced by $\mu(\ ,t)$] 139_4 Δ^r repl. by Δ_r
142_3 $F_{\tilde{w}_k} d\tilde{w}_k + F_{\tilde{w}_k} d\tilde{w}_k$ repl. by $F_{\tilde{w}_k} d\tilde{w}_k + \overline{F_{\tilde{w}_k}}\, d\tilde{w}_k$
143^{12} w_1 repl. by w_2 143_9 17 repl. by 18 147^{15} z D repl. by $z_0 \in D$
148^{18} <u>of</u> R_f repl. by R_f 148^{19} . repl. by <u>and</u> <u>equals</u> M.
148_{11} R_f repl. by R_{f*} 150^{10} $Q[f]$ repl. by $Q[\tilde{f}]$
154_7 $t_n \overline{\varphi_n} / \varphi_n$ repl. by $t_n \overline{\varphi_n} / |\varphi_n|$ 156_7 integral repl. by integrand
159_5 $1/\exp(U + iV)$ repl. by $1/\exp i(U + iV)$
160_1 <u>functions</u> repl. by <u>function</u> 173^{13} $g \circ f\ z$ repl. by $g \circ f(z)$

Vol. 817: L. Gerritzen, M. van der Put, Schottky Groups and Mumford Curves. VIII, 317 pages. 1980.

Vol. 818: S. Montgomery, Fixed Rings of Finite Automorphism Groups of Associative Rings. VII, 126 pages. 1980.

Vol. 819: Global Theory of Dynamical Systems. Proceedings, 1979. Edited by Z. Nitecki and C. Robinson. IX, 499 pages. 1980.

Vol. 820: W. Abikoff, The Real Analytic Theory of Teichmüller Space. VII, 144 pages. 1980.

Vol. 821: Statistique non Paramétrique Asymptotique. Proceedings, 1979. Edited by J.-P. Raoult. VII, 175 pages. 1980.

Vol. 822: Séminaire Pierre Lelong–Henri Skoda, (Analyse) Années 1978/79. Proceedings. Edited by P. Lelong et H. Skoda. VIII, 356 pages, 1980.

Vol. 823: J. Král, Integral Operators in Potential Theory. III, 171 pages. 1980.

Vol. 824: D. Frank Hsu, Cyclic Neofields and Combinatorial Designs. VI, 230 pages. 1980.

Vol. 825: Ring Theory, Antwerp 1980. Proceedings. Edited by F. van Oystaeyen. VII, 209 pages. 1980.

Vol. 826: Ph. G. Ciarlet et P. Rabier, Les Equations de von Kármán. VI, 181 pages. 1980.

Vol. 827: Ordinary and Partial Differential Equations. Proceedings, 1978. Edited by W. N. Everitt. XVI, 271 pages. 1980.

Vol. 828: Probability Theory on Vector Spaces II. Proceedings, 1979. Edited by A. Weron. XIII, 324 pages. 1980.

Vol. 829: Combinatorial Mathematics VII. Proceedings, 1979. Edited by R. W. Robinson et al.. X, 256 pages. 1980.

Vol. 830: J. A. Green, Polynomial Representations of GL_n. VI, 118 pages. 1980.

Vol. 831: Representation Theory I. Proceedings, 1979. Edited by V. Dlab and P. Gabriel. XIV, 373 pages. 1980.

Vol. 832: Representation Theory II. Proceedings, 1979. Edited by V. Dlab and P. Gabriel. XIV, 673 pages. 1980.

Vol. 833: Th. Jeulin, Semi-Martingales et Grossissement d'une Filtration. IX, 142 Seiten. 1980.

Vol. 834: Model Theory of Algebra and Arithmetic. Proceedings, 1979. Edited by L. Pacholski, J. Wierzejewski, and A. J. Wilkie. VI, 410 pages. 1980.

Vol. 835: H. Zieschang, E. Vogt and H.-D. Coldewey, Surfaces and Planar Discontinuous Groups. X, 334 pages. 1980.

Vol. 836: Differential Geometrical Methods in Mathematical Physics. Proceedings, 1979. Edited by P. L. García, A. Pérez-Rendón, and J. M. Souriau. XII, 538 pages. 1980.

Vol. 837: J. Meixner, F. W. Schäfke and G. Wolf, Mathieu Functions and Spheroidal Functions and their Mathematical Foundations Further Studies. VII, 126 pages. 1980.

Vol. 838: Global Differential Geometry and Global Analysis. Proceedings 1979. Edited by D. Ferus et al. XI, 299 pages. 1981.

Vol. 839: Cabal Seminar 77 – 79. Proceedings. Edited by A. S. Kechris, D. A. Martin and Y. N. Moschovakis. V, 274 pages. 1981.

Vol. 840: D. Henry, Geometric Theory of Semilinear Parabolic Equations. IV, 348 pages. 1981.

Vol. 841: A. Haraux, Nonlinear Evolution Equations- Global Behaviour of Solutions. XII, 313 pages. 1981.

Vol. 842: Séminaire Bourbaki vol. 1979/80. Exposés 543–560. IV, 317 pages. 1981.

Vol. 843: Functional Analysis, Holomorphy, and Approximation Theory. Proceedings. Edited by S. Machado. VI, 636 pages. 1981.

Vol. 844: Groupe de Brauer. Proceedings. Edited by M. Kervaire and M. Ojanguren. VII, 274 pages. 1981.

Vol. 845: A. Tannenbaum, Invariance and System Theory: Algebraic and Geometric Aspects. X, 161 pages. 1981.

Vol. 846: Ordinary and Partial Differential Equations, Proceedings. Edited by W. N. Everitt and B. D. Sleeman. XIV, 384 pages. 1981.

Vol. 847: U. Koschorke, Vector Fields and Other Vector Bundle Morphisms – A Singularity Approach. IV, 304 pages. 1981.

Vol. 848: Algebra, Carbondale 1980. Proceedings. Ed. by R. K. Amayo. VI, 298 pages. 1981.

Vol. 849: P. Major, Multiple Wiener-Itô Integrals. VII, 127 pages. 1981.

Vol. 850: Séminaire de Probabilités XV. 1979/80. Avec table générale des exposés de 1966/67 à 1978/79. Edited by J. Azéma and M. Yor. IV, 704 pages. 1981.

Vol. 851: Stochastic Integrals. Proceedings, 1980. Edited by D. Williams. IX, 540 pages. 1981.

Vol. 852: L. Schwartz, Geometry and Probability in Banach Spaces. X, 101 pages. 1981.

Vol. 853: N. Boboc, G. Bucur, A. Cornea, Order and Convexity in Potential Theory: H-Cones. IV, 286 pages. 1981.

Vol. 854: Algebraic K-Theory. Evanston 1980. Proceedings. Edited by E. M. Friedlander and M. R. Stein. V, 517 pages. 1981.

Vol. 855: Semigroups. Proceedings 1978. Edited by H. Jürgensen, M. Petrich and H. J. Weinert. V, 221 pages. 1981.

Vol. 856: R. Lascar, Propagation des Singularités des Solutions d'Equations Pseudo-Différentielles à Caractéristiques de Multiplicités Variables. VIII, 237 pages. 1981.

Vol. 857: M. Miyanishi. Non-complete Algebraic Surfaces. XVIII, 244 pages. 1981.

Vol. 858: E. A. Coddington, H. S. V. de Snoo: Regular Boundary Value Problems Associated with Pairs of Ordinary Differential Expressions. V, 225 pages. 1981.

Vol. 859: Logic Year 1979–80. Proceedings. Edited by M. Lerman, J. Schmerl and R. Soare. VIII, 326 pages. 1981.

Vol. 860: Probability in Banach Spaces III. Proceedings, 1980. Edited by A. Beck. VI, 329 pages. 1981.

Vol. 861: Analytical Methods in Probability Theory. Proceedings 1980. Edited by D. Dugué, E. Lukacs, V. K. Rohatgi. X, 183 pages. 1981.

Vol. 862: Algebraic Geometry. Proceedings 1980. Edited by A. Libgober and P. Wagreich. V, 281 pages. 1981.

Vol. 863: Processus Aléatoires à Deux Indices. Proceedings, 1980. Edited by H. Korezlioglu, G. Mazziotto and J. Szpirglas. V, 274 pages. 1981.

Vol. 864: Complex Analysis and Spectral Theory. Proceedings, 1979/80. Edited by V. P. Havin and N. K. Nikol'skii. VI, 480 pages. 1981.

Vol. 865: R. W. Bruggeman, Fourier Coefficients of Automorphic Forms. III, 201 pages. 1981.

Vol. 866: J.-M. Bismut, Mécanique Aléatoire. XVI, 563 pages. 1981.

Vol. 867: Seminaire d'Algèbre Paul Dubreil et Marie-Paule Malliavin. Proceedings, 1980. Edited by M.-P. Malliavin. V, 476 pages. 1981.

Vol. 868: Surfaces Algébriques. Proceedings 1976–78. Edited by J. Giraud, L. Illusie et M. Raynaud. V, 314 pages. 1981.

Vol. 869: A. V. Zelevinsky, Representations of Finite Classical Groups. IV, 184 pages. 1981.

Vol. 870: Shape Theory and Geometric Topology. Proceedings, 1981. Edited by S. Mardešić and J. Segal. V, 265 pages. 1981.

Vol. 871: Continuous Lattices. Proceedings, 1979. Edited by B. Banaschewski and R.-E. Hoffmann. X, 413 pages. 1981.

Vol. 872: Set Theory and Model Theory. Proceedings, 1979. Edited by R. B. Jensen and A. Prestel. V, 174 pages. 1981.

Vol. 873: Constructive Mathematics, Proceedings, 1980. Edited by F. Richman. VII, 347 pages. 1981.

Vol. 874: Abelian Group Theory. Proceedings, 1981. Edited by R. Göbel and E. Walker. XXI, 447 pages. 1981.

Vol. 875: H. Zieschang, Finite Groups of Mapping Classes of Surfaces. VIII, 340 pages. 1981.

Vol. 876: J. P. Bickel, N. El Karoui and M. Yor. Ecole d'Eté de Probabilités de Saint-Flour IX – 1979. Edited by P. L. Hennequin. XI, 280 pages. 1981.

Vol. 877: J. Erven, B.-J. Falkowski, Low Order Cohomology and Applications. VI, 126 pages. 1981.

Vol. 878: Numerical Solution of Nonlinear Equations. Proceedings, 1980. Edited by E. L. Allgower, K. Glashoff, and H.-O. Peitgen. XIV, 440 pages. 1981.

Vol. 879: V. V. Sazonov, Normal Approximation – Some Recent Advances. VII, 105 pages. 1981.

Vol. 880: Non Commutative Harmonic Analysis and Lie Groups. Proceedings, 1980. Edited by J. Carmona and M. Vergne. IV, 553 pages. 1981.

Vol. 881: R. Lutz, M. Goze, Nonstandard Analysis. XIV, 261 pages. 1981.

Vol. 882: Integral Representations and Applications. Proceedings, 1980. Edited by K. Roggenkamp. XII, 479 pages. 1981.

Vol. 883: Cylindric Set Algebras. By L. Henkin, J. D. Monk, A. Tarski, H. Andréka, and I. Németi. VII, 323 pages. 1981.

Vol. 884: Combinatorial Mathematics VIII. Proceedings, 1980. Edited by K. L. McAvaney. XIII, 359 pages. 1981.

Vol. 885: Combinatorics and Graph Theory. Edited by S. B. Rao. Proceedings, 1980. VII, 500 pages. 1981.

Vol. 886: Fixed Point Theory. Proceedings, 1980. Edited by E. Fadell and G. Fournier. XII, 511 pages. 1981.

Vol. 887: F. van Oystaeyen, A. Verschoren, Non-commutative Algebraic Geometry, VI, 404 pages. 1981.

Vol. 888: Padé Approximation and its Applications. Proceedings, 1980. Edited by M. G. de Bruin and H. van Rossum. VI, 383 pages. 1981.

Vol. 889: J. Bourgain, New Classes of $\mathcal{L}^p$-Spaces. V, 143 pages. 1981.

Vol. 890: Model Theory and Arithmetic. Proceedings, 1979/80. Edited by C. Berline, K. McAloon, and J.-P. Ressayre. VI, 306 pages. 1981.

Vol. 891: Logic Symposia, Hakone, 1979, 1980. Proceedings, 1979, 1980. Edited by G. H. Müller, G. Takeuti, and T. Tugué. XI, 394 pages. 1981.

Vol. 892: H. Cajar, Billingsley Dimension in Probability Spaces. III, 106 pages. 1981.

Vol. 893: Geometries and Groups. Proceedings. Edited by M. Aigner and D. Jungnickel. X, 250 pages. 1981.

Vol. 894: Geometry Symposium. Utrecht 1980, Proceedings. Edited by E. Looijenga, D. Siersma, and F. Takens. V, 153 pages. 1981.

Vol. 895: J.A. Hillman, Alexander Ideals of Links. V, 178 pages. 1981.

Vol. 896: B. Angéniol, Familles de Cycles Algébriques – Schéma de Chow. VI, 140 pages. 1981.

Vol. 897: W. Buchholz, S. Feferman, W. Pohlers, W. Sieg, Iterated Inductive Definitions and Subsystems of Analysis: Recent Proof-Theoretical Studies. V, 383 pages. 1981.

Vol. 898: Dynamical Systems and Turbulence, Warwick, 1980. Proceedings. Edited by D. Rand and L.-S. Young. VI, 390 pages. 1981.

Vol. 899: Analytic Number Theory. Proceedings, 1980. Edited by M.I. Knopp. X, 478 pages. 1981.

Vol. 900: P. Deligne, J. S. Milne, A. Ogus, and K.-Y. Shih, Hodge Cycles, Motives, and Shimura Varieties. V, 414 pages. 1982.

Vol. 901: Séminaire Bourbaki vol. 1980/81 Exposés 561–578. III, 299 pages. 1981.

Vol. 902: F. Dumortier, P.R. Rodrigues, and R. Roussarie, Germs of Diffeomorphisms in the Plane. IV, 197 pages. 1981.

Vol. 903: Representations of Algebras. Proceedings, 1980. Edited by M. Auslander and E. Lluis. XV, 371 pages. 1981.

Vol. 904: K. Donner, Extension of Positive Operators and Korovkin Theorems. XII, 182 pages. 1982.

Vol. 905: Differential Geometric Methods in Mathematical Physics. Proceedings, 1980. Edited by H.-D. Doebner, S.J. Andersson, and H.R. Petry. VI, 309 pages. 1982.

Vol. 906: Séminaire de Théorie du Potentiel, Paris, No. 6. Proceedings. Edité par F. Hirsch et G. Mokobodzki. IV, 328 pages. 1982.

Vol. 907: P. Schenzel, Dualisierende Komplexe in der lokalen Algebra und Buchsbaum-Ringe. VII, 161 Seiten. 1982.

Vol. 908: Harmonic Analysis. Proceedings, 1981. Edited by F. Ricci and G. Weiss. V, 325 pages. 1982.

Vol. 909: Numerical Analysis. Proceedings, 1981. Edited by J.P. Hennart. VII, 247 pages. 1982.

Vol. 910: S.S. Abhyankar, Weighted Expansions for Canonical Desingularization. VII, 236 pages. 1982.

Vol. 911: O.G. Jørsboe, L. Mejlbro, The Carleson-Hunt Theorem on Fourier Series. IV, 123 pages. 1982.

Vol. 912: Numerical Analysis. Proceedings, 1981. Edited by G. A. Watson. XIII, 245 pages. 1982.

Vol. 913: O. Tammi, Extremum Problems for Bounded Univalent Functions II. VI, 168 pages. 1982.

Vol. 914: M. L. Warshauer, The Witt Group of Degree k Maps and Asymmetric Inner Product Spaces. IV, 269 pages. 1982.

Vol. 915: Categorical Aspects of Topology and Analysis. Proceedings, 1981. Edited by B. Banaschewski. XI, 385 pages. 1982.

Vol. 916: K.-U. Grusa, Zweidimensionale, interpolierende Lg-Splines und ihre Anwendungen. VIII, 238 Seiten. 1982.

Vol. 917: Brauer Groups in Ring Theory and Algebraic Geometry. Proceedings, 1981. Edited by F. van Oystaeyen and A. Verschoren. VIII, 300 pages. 1982.

Vol. 918: Z. Semadeni, Schauder Bases in Banach Spaces of Continuous Functions. V, 136 pages. 1982.

Vol. 919: Séminaire Pierre Lelong – Henri Skoda (Analyse) Années 1980/81 et Colloque de Wimereux, Mai 1981. Proceedings. Edité par P. Lelong et H. Skoda. VII, 383 pages. 1982.

Vol. 920: Séminaire de Probabilités XVI, 1980/81. Proceedings. Edité par J. Azéma et M. Yor. V, 622 pages. 1982.

Vol. 921: Séminaire de Probabilités XVI, 1980/81. Supplément: Géométrie Différentielle Stochastique. Proceedings. Edité par J. Azéma et M. Yor. III, 285 pages. 1982.

Vol. 922: B. Dacorogna, Weak Continuity and Weak Lower Semicontinuity of Non-Linear Functionals. V, 120 pages. 1982.

Vol. 923: Functional Analysis in Markov Processes. Proceedings, 1981. Edited by M. Fukushima. V, 307 pages. 1982.

Vol. 924: Séminaire d'Algèbre Paul Dubreil et Marie-Paule Malliavin. Proceedings, 1981. Edité par M.-P. Malliavin. V, 461 pages. 1982.

Vol. 925: The Riemann Problem, Complete Integrability and Arithmetic Applications. Proceedings, 1979-1980. Edited by D. Chudnovsky and G. Chudnovsky. VI, 373 pages. 1982.

Vol. 926: Geometric Techniques in Gauge Theories. Proceedings, 1981. Edited by R. Martini and E.M.de Jager. IX, 219 pages. 1982.